AURALITY

ANA MARÍA OCHOA GAUTIER

SIGN, STORAGE, TRANSMISSION • *A series edited by Jonathan Sterne and Lisa Gitelman*

AURALITY

LISTENING AND KNOWLEDGE IN

NINETEENTH-CENTURY COLOMBIA

DUKE UNIVERSITY PRESS · DURHAM AND LONDON · 2014

Printed in the United States of America on acid-free paper ∞
Designed by Courtney Leigh Baker
Typeset in Arno Pro and Trade Gothic by Graphic Composition, Inc., Bogart, GA

Library of Congress Cataloging-in-Publication Data
Ochoa Gautier, Ana María.
Aurality : listening and knowledge in nineteenth-century Colombia /
Ana María Ochoa Gautier.
pages cm—(Sign, storage, transmission)
Includes bibliographical references and index.
ISBN 978-0-8223-5736-0 (cloth : alk. paper)
ISBN 978-0-8223-5751-3 (pbk : alk. paper)
1. Oral communication—Colombia—History—19th century.
2. Listening—Social aspects—Colombia—History—19th century.
3. Voice—Social aspects—Colombia—History—19th century.
4. Sound—Social aspects—Colombia—History—19th century.
I. Title. II. Series: Sign, storage, transmission.
P95.43.C7O24 2014
302.2′24209861—dc23
2014006960

Cover art: Olga de Amaral, *Entre rios I*, 2010. Linen, gesso, acrylic, and gold leaf. 185 × 170 cm.
Courtesy of the artist. Photograph © Diego Amaral.

A María Eugenia Londoño y Jesús Martín-Barbero,
A Richard Bauman y Beverly Stoeltje • Para Alina y Bernardo

CONTENTS

This book was not initially about listening, the voice, or the nineteenth century. The original project for the book was an inquiry into the intellectual history of studies of popular music in Colombia from the 1930s to the 1970s. I was especially interested in the correlation between music and the literary, since the entanglement of the sonorous in different forms of narrative has been central to Latin American musical thought, and many of its foundational figures have been more widely recognized as writers than as music scholars. Also, the field of folkloristics was historically shaped in such a way that the study of the musical in the popular was often undertaken as an intellectual and political endeavor that crossed the anthropological, the musical, and the literary with multiple practices of cultural policy.

But when I went to the archives of the National Radio in Colombia (Radio Nacional de Colombia) in 2008 to listen to the radio programs Colombian folklorists had done throughout their career, I realized that programs on folklore from the 1940s to the 1980s used what was initially a vanguard technology, the radio, to promote a conservative listening pedagogy that constantly cited nineteenth century Colombian sources. I then turned to the nineteenth century archive to begin an initial exploration of the question of why such a listening pedagogy had been so persistent, shaping many of the ideas about the notions of *música popular*. To my surprise, I found an archive full of listening practices. As I found and read dispersed materials on such listening practices in the nineteenth century, a sonorous written archive I had never suspected was there began to take shape. As that archive took form, the topic of this book gradually changed from the intellectual history of twentieth century folklorists to the role of listening to different sounds considered "voices" in shaping the notions of nature and culture, so central to understandings of personhood and alterity that imbue the popular in Latin America.

This book took shape in the midst of conversations on popular music going back and forth between colleagues in Colombia, the United States and other Latin American countries, particularly Brazil and Argentina. I am deeply indebted to the possibility of such conversations across different places and the way they transformed my own thinking. I also did the research and writing for this book, as I learned to teach in the United States and in the contrasts between working in cultural policy in Colombia, where I did not have an academic job in a university, and in the academy in the United States. This book then, emerged in a period of big transformations in my life and was formed by questions that took shape with the people and in the events that accompanied its own emergence. It helped me mediate many transformations, celebrate different conversations and their ramifications for further thought, and provided a thread of inquiry into the many incommensurables and closed avenues that also shaped its final form. As such, the book seems to name for me more a particular moment of my life than a particular work. This book then is in gratitude to the many people who have been present in this moment across so many places and changes, populating it in different ways. The people and institutions I name here then are the ones that are most immediate to the emergence of the physical book, especially its final stage, than to the general intellectual and affective processes that made it possible.

The research for this book was made possible by a generous grant from the Guggenheim Foundation, and by the support of Columbia University and New York University. Some of the ideas in this book were discussed in presentations in several conferences and I benefited tremendously from those discussions. Some of those were the meeting of the Associação Brasileira de Etnomusicologia in Maceió, Brazil in 2008, the meetings of the Asociación Argentina de Musicología in 2008 and of IASPM–Latin America in Córdoba in 2012, and several meetings of the Society for Ethnomusicology in the United States. Some of the topics in this paper were originally presented at conferences in Columbia University, Duke University, New York University, Universidad Nacional del Rosario, Universidad Nacional de Córdoba, Universidad de Buenos Aires, Universidad de Cartagena, Universidade Federal do Rio de Janeiro, Universidad de los Andes, Universidad Javeriana, University of California, San Diego, Universidad Andina Simón Bolívar, University of California, Los Angeles and in the working group on music research and policy in Colombia promoted by the Ministry of Culture.

My colleagues from the Department of Music at Columbia University have offered generous support at different moments. I especially want to thank fellow ethnomusicologists Aaron Fox, Ellen Gray, and Chris Washburne for their generosity these past years. Samuel Araujo, Dora Brausin, Carolina Botero, Charles Briggs, Juan Calvi, Alessandra Ciucci, Arturo de la Pava, Adriana Escobar, Arturo Escobar, Jorge Franco, Miguel García, Julio Gaitán, Gabriel Gómez, Adriana Henao, Carlos Miñana, Rubén López Cano, Alejandro Mantilla, Pablo Mora, Cláudia Neiva de Matos, Jorge Mario Múnera, Mauricio Pardo, Lucía Pulido, Jaime Quevedo, Eduardo Restrepo, Omar Romero, María Alejandra Sanz, Carolina Santamaría, Steven Shaviro, Jessica Schwartz, Christian Spencer, Gavin Steingo, Jonathan Sterne, Alejandro Tobón, Elizabeth Travassos, Martha Ulhoa, María Victoria Uribe, Miriam Vergara, and Leonardo Waisman listened and provided silent or explicit advice I sometimes heeded, sometimes not.

Farzanneh Hemmasi, Morgan Luker, Amanda Minks, David Novak, Matt Sakakeeny and Anna Stirr initially motivated me to write this book in English and became my imaginary audience when I struggled to find words to translate, from Spanish to English, Latin American concepts that are seldom discussed in Departments of Music and music scholarship. Louise Meintjes and Jairo Moreno have been key tricksters in helping me figure out what road to take. Julio Ramos arrived at the final stages of this book and provided much needed conversation, advice, encouragement, and support in the long process of giving it closure.

Some of the ideas in this book coalesced in courses I taught at Columbia University throughout the years on music, literature, and critical thinking in Latin America, on ethnomusicological thought in the nineteenth century, on music and politics, and on sound, the sacred, and the secular. Also important were the two courses on music and politics and on music and critical thinking in Latin America taught in the Masters in Cultural Studies at the Departamento de Lenguajes y Estudios Socioculturales in the Universidad de los Andes in 2011. I want to thank the students who participated in those courses for insightful discussions that helped me shape and rethink the material.

Samuel Araujo, Licia Fiol-Matta, Aaron Fox, Gavin Steingo, Morgan Luker, Louise Meintjes, Jairo Moreno, Thomas Porcello, Julio Ramos, David Samuels and Steven Shaviro read fragments of the manuscript at different moments and provided a much needed sounding board to ideas that needed to breathe in other people's minds. Morgan Luker read an initial draft of this book several years ago and was extremely helpful in helping me understand sentence structure in English. Ken Wissoker has been an excellent editor discussing ideas, doubts and guiding me in the whole process of producing a book in the university publishing system in the United States. Susan Albury and Jillian O'Connor have been incredibly helpful in giving advice, providing insights and responses to different editorial questions. César Colón-Montijo and Laura Jordán provided invaluable help in double checking the correspondence between references and bibliography. As they say in English, the responsibility for the form and content of the book is, of course, mine.

Katie Aiken, Yenny Alexandra Chaverra, Lucila Escamilla, Pablo Mora, Ani Yadira Niño, Jaime Quevedo and Omar Romero helped me document, organize, and double check references in my dispersed archive at different moments of the research and writing process. I thank the librarians at the Biblioteca Nacional de Colombia, Museo Nacional de Colombia, Fonoteca de la Radio Nacional, Centro de Documentación Musical of the Biblioteca Nacional de Colombia, Biblioteca Luis Angel Arango, Columbia University, and New York Public Library for their invaluable help in locating books, radio programs, and original manuscripts.

I have been fortunate to have had good teachers who have provided an effective intellectual environment that nurtured my interest in genealogies of thought and intellectual histories. This book is dedicated to my teachers Maria Eugenia Londoño and Jesús Martín-Barbero in Colombia, and Richard Bauman and Beverly Stoeltje in the United States. I first met María Eugenia Londoño when I was seven years old when she became my next door neighbor and my piano teacher. She has been instrumental in developing modes of music research associated to action-participation in Colombia, a generous teacher, and close friend. Even though our ideas about specific topics are frequently different, no other scholar or author has shaped my thought on politics and music as much as she has. Jesús Martín-Barbero provided discussion, intellectual support and friendship, especially in the years I lived and worked in Colombia after finishing my doctoral degree. His modes of weaving genealogical thinking and questions about popular culture were crucial for shaping my own interest on archaeologies and genealogies of the popular.

I never published my PhD dissertation due to circumstances whose description and explanation are beyond this text. But that means the presence of many ideas developed initially during the years of my doctoral degree stayed with me as lines of inquiry that coalesced in this book in a different and new form. Richard Bauman encouraged bringing together questions shaped by changing ideas on the popular that were taking place in Latin America in the 1990s with changing ideas about folklore and sociolinguistics that were also taking shape in the same period in the United States. I thank him for seeing and pointing out avenues of inquiry that eventually became central to this project. Beverly Stoeltje patiently and with good humor taught me about mediations between theory, expressive culture, ethnographic practice, and personal desires and quests. Her own borderland modes of bridging North and South have been central to the theoretical consolidation of my own work.

Alina and Bernardo, my parents, are key intellectual figures in my own formation. I thank my mom one day for bringing home a copy of Candelario

Obeso's *Cantos Populares de mi Tierra* that she bought in a stand at a super-market in Medellín. She said that she thought the book might interest me. When she realized I was writing about philologists, she gave me her own very used 1951 copy of Marroquín's *Manual on Orthography*. She has always linked affect, research, and university teaching in ways that are still baffling to me. I wish I had her capacity for analyzing situations and almost instantly respond-ing with the appropriate words and gesture. Her mode of wisdom and intel-lectual engagement is something I slowly learn. Bernardo has taught me that histories of knowledge and being are always about politics and that politics are about struggles, negotiation, debate, and controversy. His willingness to analyze, think, speak, and ask (not necessarily in that order) in the midst of very difficult and controversial political moments in the history of our lives in Medellín has been a crucial example. Both taught me, in different ways, the craft and art in reason and the reason and art in craft-making. This book is for both of them in loving celebration of their lives. I thank Augusto, Juan, Jennifer, Beatriz, Bernardo, Lucas, and Sara for their presence as we learn to shape lives between North and South.

THE EAR AND THE VOICE IN THE LETTERED CITY'S
GEOPHYSICAL HISTORY

On January 5, 1884, Colombian philologist Rufino José Cuervo (Bogotá 1844–Paris 1911) wrote a letter from Paris to Miguel Antonio Caro (1845–1909), his fellow grammarian and president of Colombia from 1892 to 1898, who was then residing in Bogotá: "Do you know if somebody has thought about collecting [in Colombia] housemaid tales such as those collected by Grimm and Andersen?" (Cuervo 1978, 111). What was initially posed as a question soon became an affirmation. The apparent lack of documentation of a collected folk corpus has often led to the assertion that in the nineteenth century there were very few studies of folk expressions in Colombia. Gustavo Otero Muñoz, for example, wrote in 1928 in his book *La literatura colonial de Colombia seguida de un cancionerillo popular* [Colombian colonial literature followed by a small songbook]: "The Republic of Colombia is behind in the work that corresponds to it, as a civilized country, namely, that of contributing its fragment of truth regarding the common heritage of the species, formed by the science that Grimm, Max Muller, Bopp and so many other wise men have glorified, searching in the traditions of each regional folklore, the bond that brings into relation the different religions and languages of the peoples" (Otero Muñoz 1928, 241).

This idea of the lack of serious studies of local expressive cultures (including music) has persisted well into the present.[1] Whether through absence of documentation or through the use of inappropriate methodologies in the study of local expressive culture, this seeming lack has acquired a foundational character, the aura of a national truth that hauntingly returns during different historical moments to account for different aspects of Colombia's conflictive history.

The perception and reiteration of this void since the nineteenth century is even more surprising given the fact that Colombia was a key site of global botanical and geographic scientific expeditions in the eighteenth and nineteenth centuries as well as a country of internationally renowned philologists, particularly in the late nineteenth century. The Botanical Expedition of the Kingdom of New Granada began in 1783, lasted twenty-five years and was crucial in constructing the idea of New Granada as botanically exuberant and for creating the field of natural sciences in the country (Nieto Olarte 2000, 2006). The Geographical Commission's (*Comisión Corográfica*) primary work of description and mapping of the provinces of the country, which took place between 1850 and 1859, produced a wealth of maps and ethnographic documentation, and was important for the foundation of such disciplines as geography, engineering, and ethnology (Sánchez 1999). During the second half of the nineteenth century, Colombian philologists, poets, and writers from different regions of the country developed a corpus of written genres such as textual annotations on maps, customs sketches, enlightened travel writing, poetry, and novels imbued with local idioms, detailed philological analysis of local language usage, annotated reeditions of colonial indigenous grammars, histories of Colombian literature, among others, that included the depiction, theorization, or usage of local language and expressive practices. Moreover, philology became a highly politically charged discipline in the late nineteenth century, when a series of "grammarian presidents," philologists who came to presidential power and who were deeply concerned with the proper use of language, brought the full import of their knowledge into the national design of jurisprudence, education, and religious affairs.

But none of these dispersed disciplinary concerns with the local have historically counted as proper "folklore collections." If, during the eighteenth and nineteenth centuries, the sciences of nature and those of language and expressive culture were not seen as totally separate, but as part and parcel of the epistemological endeavor of building a corpus of knowledge about the nation at a historical moment when such an endeavor was an urgent political necessity, then it seems awkward that there was such a distance between the wealth of information generated by the botanical and cartographic expeditions, the abundance of philological and poetic texts, and the apparent void in documenting local aural expressions. Even more so if we consider that due to its varied geography, and following an Ibero-American lineage of identifying the Andes with paradisiacal excess, during this period Colombia was perceived as a country of botanical abundance and unlimited economic potential that could produce any type of natural commodity found on the globe (Cañizares-Esguerra 2006; Nieto Olarte 2000). Rather than a problem to be solved, I see this disjuncture as a key site for understanding how the idea of a valid aural expressive genre was constituted depending on the listening practices or "audile techniques" (Sterne 2003) through which it was constituted. What is revealed by such a disjuncture is that many of the acoustic dimensions of the colonial and early postcolonial archive are not presented to us as discrete, transcribed works or as forms neatly packaged into identifiable genres (Tomlinson 2007). They are instead dispersed into different types of written inscriptions that transduce different audile techniques into specific legible sound objects of expressive culture. This book is about ontologies and epistemologies of the acoustic, particularly the voice, produced by and enmeshed in different audile techniques, in which sound appears simultaneously as a force that constitutes the world and a medium for constructing knowledge about it. Voice was ambiguously located between "nature" and "culture," and thus was central for shaping what those terms meant in this historical period. I explore how listening practices were crucial in determining how the voice was understood and what counted as a proper form of voicing and cultural expression for different peoples in Colombia at a historical moment when the colonial itself had to be reformulated as a postcolonial politics of an independent nation.

An acoustically tuned exploration of the written archive reveals that the documentation of local expressive aural practices was entangled in what was then understood as natural and civil histories as well as by emergent creative practices in the fine arts. It is thus difficult in this period to find separate folklore collections neatly packaged and understood as such. But there was abundant discussion on the sounds of Colombia's many different peoples, nonhuman animals, and entities of nature–rivers, volcanoes, the wind. The full import of different practices of rationalization and artistic creativity was used in making sense of such listenings, simultaneously producing knowledge about local soundings and a "reorganization" (Rancière *dixit*) of how the senses were perceived, felt, understood, and used. In this book I seek to explore how different practices of listening led to the inscription in writing of local aural expressive genres as well as to an enlightened cultivation of hearing that were crucial to the development of concepts about "local culture" and "local nature" that often persist to our days. In the process what emerged was not only a dispersed corpus of ideas of how to think about local creative cultures and about local nature but also a "refunctionalization of the ear" (Steege 2012) and its relation to the voice.

In Latin America and the Caribbean, different moments and processes of aural perception and sonic recontextualization have always been accompanied by an intense debate about the meaning of sonic localism and temporality and its place in history. In his book *The Lettered City*, Angel Rama saw the written word, concentrated in the cities, the sites of political administration, as constitutive of a highly unequal public sphere that took shape in the hands of lettered elites that to him resembled a "priestly caste . . . that enjoyed dominion over the subsidiary absolutes of the universe of signs" ([1984] 1996, 16). But in this book I argue that Latin America was simultaneously and just as importantly constituted by audile techniques cultivated by both the lettered elite and peoples historically considered "nonliterate," giving rise to the types of questions and relations that the worlding of sound enables. Lettered elites constantly encountered sounding and listening practices that differed from their own: vocalities that seemed out of tune, difficult to classify as either language or song, improper Spanish accents that did not conform to a supposed norm, sounds of indigenous languages for which there were no signs in the Spanish alphabet, an abundance of noises or "voices" coming from natural entities that seemed to overwhelm the senses. In the process of inscribing such listenings into writing, the lettered men (and it was mostly men) of the period simultaneously described them, judged them, and theorized them. And while some were keen to rein in what seemed like a disordered acoustic abundance into a descriptive and normative standard that allowed for the proper identification of an ordered "nature" and "culture," others sought to enhance the relevance of such acoustic multiplicity by reveling, often in contradictory ways, on the significance of such sensorial exuberance.

In the midst of such processes of recording sound in writing, what emerges is not only the possibility of exploring how the lettered elites conceived of sounds. By reading the archive against the grain, it is possible to speculate how indigenous peoples, Afrodescendants, and mestizos also conceived of such vocalizations. Thus, rather than seeing the nineteenth century (or the colony) solely as the site of constitution of Western theories about the other, I prefer to understand it as a contested site of different acoustic practices, a layering of contrastive listenings and their cosmological underpinnings. The different practices through which such listenings have been historically inscribed, in bodies, on stone, on skin, and on paper, through rituals and through writing, are, to be sure, marked by highly unequal power in the constitution of the public sphere. But that does not mean that their significance disappeared or was completely erased. Rather they had to be accounted for, even if to deny their singularity while acknowledging them as resources for the distinction between popular and fine arts, between an improper pronunciation and a good one, or, frequently, to unsettle the taken for granted tenets of the disciplining of sound. Thus, in this book, the aural is not the other of the lettered city but rather a formation and a force that seeps through its crevices demanding the attention of its listeners, sometimes questioning and sometimes upholding, explicitly or implicitly, its very foundations.

One of the elements that emerged as I explored the archive was that ideas about sound, especially the voice, were central to the very definition of life. In this book I explore how persistent underlying understandings of the acoustic today emerged or were consolidated during the early postcolonial period, especially regarding the way "local sounds" of different entities and of peoples were understood as "voices." I am particularly interested in what the endurance of some of the often unacknowledged or taken for granted ideas regarding local expressive cultures, especially in a "globalized" world and one increasingly marked by the condition of displacement, is supposed to invoke, provoke, and incarnate for different peoples and the different formulations of what they are supposed to acoustically embody and be. Understandings of the significance of local linguistic accents, of the relations between body and voice, of the seemingly contradictory politics of education of the ear and the voice of "the people" amidst the simultaneous recognition of a diversity of listening and vocal practices were crystallized or (re)formulated for the purposes of the nation in the nineteenth century, even when deeply embedded in colonial history's lengthy lineages of the global/local constitution of knowledge and of ideas of personhood. The voice, especially, was understood by Creoles and European colonizers as a fundamental means to distinguish between the human and nonhuman in order to "direct the human animal in its becoming man" (Ludueña 2010, 13). Thus, the relation between the ear, the voice, and the understanding of life emerged as a particularly intriguing dimension of the politics of what a "local" expressive culture was supposed to be or become and who could incarnate it. The question of how to distinguish between human and nonhuman sounds became particularly important in a colonial context in which the question of such a boundary troubled, in different ways and for different reasons, the many peoples that originally populated, willingly came, or were forcefully brought to the Americas. And in the formulation of such a question the spectral figuration of the voice and of the acoustic as an invisible yet highly perceptible and profoundly felt (im)materiality, which hovers between live entities and the world, became particularly important.

This book is a contribution to the intellectual history of listening, a growing field in the past two decades. The current scholarly trend of searching for traces of the aural in the literary (Picker 2003; Lienhard [1990] 2011), for the sound of the voice in different historical contexts and vocal genres (Abbate 1991; B. Smith 1999; Connor 2000), the attention to different "new" technologies that existed prior to the invention of the phonograph (Gitelman 1999, 2006), the trace left by different genres of inscription on the work of music making (Szendy 2008; Johnson 1995), the critical work on the philosophical grammar of vocality and writing (Derrida [1974] 1997; Cavarero 2005; Cournut 1974), the study of discourses and practices of the aural in fields such as physiology and acoustics that surrounded the invention of sound machines (Sterne 2003; Brady 1999; Steege 2012), and the search for how specific historical periods prior to the emergence of mechanical sound reproduction sounded (Smith 2004; Rath 2003; Corbin 1998), attests to the recent scholarly recognition of these historical practices of legible aural inscription. Such work is considered part of a general auditory turn in critical scholarship, one which explores "the increasing significance of the acoustic as simultaneously a site for analysis, a medium for aesthetic engagement, and a model for theorization" (Drobnick 2004, 10).

This recent "auditory turn" in critical theory is giving rise today to the increased formalization of sound studies (Sterne 2012a). Questions about how such a turn is historically traced, what fields of sound studies are privileged in tracing such a genealogy, and who the pioneering figures are differ from one scholar to another (Sterne 2012a; Feld 1996; Szendy 2009). But another perhaps more central issue emerges here. As historical work on audition intensifies, the question is raised as to whether the deaf ears of history were those of an epochal moment when the gaze was privileged above all other senses in the West (Jay 1993), or whether listening practices were always there, hidden by the fact that the traces left by audibility are enmeshed with different practices, a listening to be found in the nooks and crannies of history, dispersed across several fields and sites of knowledge and sound inscription. For Peter Szendy, the "critical force of listening" has often been "restrained and denied" (2008, 34). Listening also appears as hidden behind other auras. For example, studies of "orality" tend to concentrate on theorizations of its literary dimensions, described as the other of writing rather than according to its own specificities, and its acoustic dimensions are often subsumed under other linguistic elements (Feld et al. [2004] 2006). Also, for Veit Erlmann, colonial and postcolonial studies have tended to privilege the gaze, and the history of sound studies is primarily on historical works in Euro-American contexts (2004). But such privileging of the gaze is increasingly questioned by rethinking in the history of the senses.[2] Moreover, a long Latin American lineage of interrelating oral and written texts has been central to rethinking the history of indigenous texts in the formation of the literary (Lienhard [1990] 2011; Sa 2004). This is part of a broader history of the gaze, print, the "oral" and the lettered word as central to the insertion of the region into the global construction of modern capitalism (Franco 2002; González Echeverría 1990; Pratt 1992; Rama [1984] 1996; Ramos [1989] 2003). This book builds on such work but inverts the emphasis on the relation between the *written text* and the *mouth* (implied by the idea of the oral) by exploring how the uses of the *ear* in relation to the *voice* imbued the *technology of writing* with the traces and excesses of the acoustic.

My own work on nineteenth-century practices of listening in the midst of the transformation of colonial New Granada[3] to national independence from Spain, seeks to contribute to the historical scholarship on the relationship between listening and the voice as part of the history of the relation between the colonial and the modern. Before the invention of sound machines, the inscription of sound took place through what Lisa Gitelman has called "legible representations of aural experience" (1999, 15). This involved not only musical notation but also words about sound and aural perception, and recognizing the different historical ways in which technologies of the legible made and still make sound circulation possible. Since the period I address is before the invention of sound reproduction machines, I necessarily work with the inscription of sounds into writing. By inscription, following Lisa Gitelman (1999, 2006), I mean the act of recording a listening into a particular technology of dissemination and transmission (in this case writing). But the inscription of sounds can also occur on the body, in different kinds of objects such as stones, waterfalls, or other entities of nature or of urban life, which are understood by different peoples as containing or indexing the sound archive (Feld 1996, 2012; Seeger 1987; Hill and Chaumeil 2011).

In this book, practices of listening and aural perceptions, descriptions, and knowledges appear dispersed across several sites of inscription: travel writing (chapter 1), novels, poems, and literary histories (chapter 2), songbooks (chapter 2), grammars (chapter 3), ethnographies and political writings on language (chapters 3 and 4), orthographies and practices of music notation (chapter 4). Listening is not a practice that is contained and readily available for the historian in one document but instead is enmeshed across multiple textualities, often mentioned in passing, and subsumed under other apparent purposes such as the literary, the grammatical, the poetic, the ritual, the disciplinary, or the ethnographic. If sound appears as particularly disseminated across different modes of inscription and textualities it is because, located between the worldly sound source from which it emanates and the ear that apprehends it, the sonorous manifests a particular form of spectrality in its acoustics.

Such a spectrality of sound also shows up in other ways. For many of the lettered men of the nineteenth century, the relation between writing as a format of inscription and listening appears as highly problematic. Angel Rama identified the power of the written word as the "autonomy of the order of signs" in Latin America, "its capacity to structure vast designs based on its own premises" ([1984] 1996, 60), a particular order of things done with written words central to the structure of governmentality in the region. And yet, the inscription of the acoustic seems to render that power as highly ambiguous. In Latin America and the Caribbean practices of "narrative transculturation" systematically seep into the written word to the point that part of the history of the region's literary aesthetics is narrated almost as a history of the practices of incorporation of the sonorous aspects of language (rendered as "popular culture") into the written realm (Rama [1984] 2007). On the one hand, writing is indeed what inscribes a proper form of listening and of vocality into the law and thus epistemologies on how the local should sound, emerge from the pen. As I explore in this book, the use of writing as a technology of inscription and dissemination determines the history of the rise of folkloristics and the politics of language and popular song in the official canon. On the other, writing just as frequently appears merely as an instrument or medium in the service of acoustic memory (Cournut 1974). As such, it is often a fallible technology, a highly limited format that renders ambiguous the relation between the voice and writing and the powers ascribed to each (Cavarero 2005; Derrida [1974] 1997). A format "names a set of rules according to which a technology can operate" (Sterne 2012, 7), and the limitations found in writing reveal the disjunctures such rules provoke. Alphabetic graphemes do not conform to linguistic sounds either of indigenous or European languages (chapters 2 and 3), staff notation is not technologically suitable for the typographic technologies of nineteenth-century Colombia (chapter 4), and the description of different types of voices through writing is at best an approximation to how they sound (chapters 1, 2, and 3). But this is not only a problem of the formats of writing. What the limits of the format make evident is that the acoustic recognition of different practices of vocalization or sounds of natural entities associated with the idea of the voice exceeds their very inscription. Through the problems presented by technological limitations, what emerges is that the ontology of the relationship between the ear and the voice exceeds its containment in a particular medium. Listening appears as the nomadic sense par excellence and the voice as highly flexible, an instrument that can be manipulated to position the relation between the body and the world in multiple ways (Weidman 2006). The politics of regimentation of the voice are also multiple and often show us how the body and the voice do not necessarily coincide (Connor 2000; Weidman 2006). To the contrary, voices have the potential to disembody themselves into objects as in ventriloquism (Connor 2000), to travel between human and nonhuman entities as when animals teach humans songs (Seeger 1987), to incarnate other worldly beings in a body of this world as in rites of "possession" (Matory 2005), or are presumed to represent an autonomous or unique individual, as in the predominant Western philosophical political tradition (Weidman 2011). Hearing voices thus frequently invokes the need to ontologically address implied questions about the cosmologies (Schmidt 2000) or the ear (Steege 2012) and the definition of life they bring forth.

The relation between the voice and the ear then implies a *zoé*, a particular notion of life that involves addressing different conceptions of the human and the boundaries between the human and nonhuman. In the colonial context of the Americas, where peoples from different places came together, such a definition of life through the voice was certainly a contested political issue. For exploring such an issue, I use the term *zoopolitics* following Fabián Ludueña (2010) in his deconstruction of the division between *bíos* and *zoé* as present in the work of Foucault and Agamben, a term he takes from Derrida. Ludueña questions the neat division between *bíos* as "something like a qualified life, and thus the more proper subject of politics while *zoé* represents, to say it in some way, a natural life originally excluded from the world of the city" (Ludueña 2010, 28).[4] The definition of "the political community of humans" (Ludueña 2010, 13) implies the definition of life to determine the boundary between the human and nonhuman. Nature is not that upon which culture builds, but rather both terms, nature and culture, are mutually constituted through the politics of life. One of the central aspects explored in each of the chapters of this book is how a zoopolitics of the voice was a political means, in this histori-cal moment of transformation from the colony to the postcolony, to redefine the relationship between the colonial and the modern.

The Aural, the Colonial, the Modern

Colonialism in Latin America and the Caribbean has generated a plurality of responses and discourses throughout its long history: "it could be argued that, at all levels, from colonial times to the present, intellectual action has been developed in an attempt to confront the traumatic effects of colonialism" (Moraña, Dussel, and Jáuregui 2008, 12). And yet, the region's rich tradition in critical readings of its own colonial history has either been largely ignored by the postcolonial debate beyond the Ibero-American world or provincialized by this very discourse. Part of the difficulty of seeing the layered critical colonial Latin American and Caribbean history for scholars working outside of the region is plain ignorance of this long debate, since the recognition of Latin American intellectual critical theory continues to be primarily tied to Spanish and Portuguese, and history departments in the Anglo-American world, and is seldom taught beyond this context, except as "local" theory on courses about Latin America in the humanities and social sciences. Also, the multiple temporal displacements of the Latin American debate as well as the diversity of positions in proposing decolonial politics throughout this long history, makes it difficult for scholars foreign to the region's debates to recognize this long legacy on critical thinking on the colonial.

The history of the American conquest and genocide as well as the history of slavery in the region have been deeply entangled with the rise of a global, capitalist modernity. This was a major change at a global scale that articulated the rise of modernity with the globalization of colonialism to the rise of capitalism (Quijano 2008). The complex ways in which global economic relations were articulated between and in different parts of the globe after the sixteenth century is giving rise to a changing critical transatlantic history that is redrawing understandings of capitalism (Tutino 2011). For Cuban anthropologist Fernando Ortiz such a history needed to be critically addressed through the interrelationship between the economic, juridical, cultural, biological, and aesthetic spheres as constituted by the changing historical politics of global economic exchange ([1940] 1987). Latin American and Caribbean colonial studies then are deeply embedded in the history of trying to account not only for the region's specific colonial history but for its imbrication in the global articulation of modernity.[5] This has broader implications even beyond the widely accepted postcolonial tenet that knowledges, economies, and histories are globally constituted in the traffic of peoples, ideas, and things between different parts of the world, mediated by unequal power relations (Chakrabarty 2000; Mignolo 2000).

As affirmed by Michel-Rolph Trouillot, one of the effects of the key role of the Caribbean region in the articulation of a global modernity was its erasure from broad anthropological significance because its overarching history was not one that could be easily construed through an isolationist nativism (1992). Something similar can be said about the contributions to the dialectic between colonial histories and decolonial thought. The region's centrality in the consolidation of modernity has often led to an erasure of its colonial history (Lomnitz 2005) and its significance for decolonial thought; an erasure that Trouillot (1995) poetically read as treading through silenced ruins that metaphorically reveal the incommensurabilities of the region's decolonial struggles in the mute speech rendered by a landscape marked by colonial architectural remnants. The deeply held tenet that Spanish colonialism and Anglo settler colonialism were highly different because one was more "rational" and the other more "irrational" has been increasingly questioned (Cañizares-Esguerra 2006a). This needs to be acknowledged as a significant aspect for rethinking the debate on colonial legacy and the formulation of indigenous studies in the North and the South. Also, the history of the colonial in Latin America and the Caribbean is a central element of the renewed significance given to the debate on the contested nature of the person and of nature. This topic has become particularly salient, partly as a response to the unprecedented devastation to the environment produced during the past 250 years or so. According to Dipesh Chakrabarty (2009), this implies a crucial redefinition of history by linking the political history of the person with the species history of the human with an urgency that it had never had before.

As both Claude Lévi-Strauss ([1955] 2012) and Eduardo Viveiros de Castro (2011) have written, the contested history of the definition of the human as a species and the political definition of the person has been a central topic of debate since early colonial times in Latin America, thus "demonstrating the necessity of taking back the 'archaeology of the human sciences' at least to the controversy of Valladolid (1550–1551), the famous debate between Las Casas and Sepúlveda about the nature of American indians" (Viveiros de Castro 2010, 28). In this book I explore the relation between the voice and the ear as a fragment of this broader history.

In chapter 1 I explore the way that vocalizations of boat rowers of the Magdalena River, or *bogas,* were heard by Creoles and Europeans and on how those same vocalizations were understood by Afrodescendants and indigenous groups in the midst of an intense process of biological mixture that characterized this region in the eighteenth century.[6] Let us recall that the global colonial archive is full of peoples who howl like animals (Tomlinson 2007) and the *bogas'* mode of vocalization was described by travelers again and again as a mode of howling comparable to the voices of different animals. Such howls were used to understand the boundary or relation between the human and the nonhuman by Western travelers in one way, and by the *bogas* and other riverine peoples from the Caribbean, in another. For Creoles and Europeans, sounding like animals was the sign of a lowly human condition, used for processes of racialization through a politics of representation. For others, such as the bogas or indigenous peoples in northern South America, the voice was not understood as that which represented their identity. Instead, the voice manifested or enabled the capacity to move between states of multiplicity or unity where a single person can envoice multiple beings and where collective singing, as in a feast, can manifest a unity in which the collective is understood as expressing the singular (Strathern 1988; Seeger 1987), where different living entities or musical instruments voice the breath of life (Hill and Chaumeil 2011) and where culture is understood "as an on-going act of creation" rather than "the distillation of a set of abstract ideals" (Guss 1989, 4).

In this chapter I also explore the work of Alexander von Humboldt in the Americas in the broader context of his role in the "Berlin Enlightenment" and the rise of an "enlightened vitalism" that challenged contemporary European mechanical understandings of natural history and aspects of the Cartesian mind-body division as definitive to the consolidation of a European Enlightenment (Reill 2005). These ideas on vitalist theories were developed among German and French scholars through intense and mostly unacknowledged exchanges with naturalists of the Americas, through writings and ideas about the region that circulated as part of the struggle between European nations over the appropriate interpretation of colonial history (Nieto Olarte 2007; Cañizares-Esguerra 2006).[7] Such an exchange was crucial for notions such as identifying a clear pitch as one of the central elements of defining a proper music in the history of musicology and comparative musicology, or the interrelationship between language and identity as a central aspect of musical nationalism as established by the Berlin Counter-Enlightenment (notably by Alexander's brother, Wilhelm), or how notions of climate as determinant factor of race, influenced what was understood as a valid music, language, and culture, in the production of racialized ideas about personhood. I question the formulation of musicology, comparative musicology (and comparative linguistics) as disciplines that were forged solely in Germany (Potter 1998), and posit them rather as disciplines that were forged through the colonial exchange of ideas and data (Bloechl 2008), that took different forms in different places, and to which the formations of knowledge that happened in and were taken from Latin America and the Caribbean were central.

That some of the major ideas about "nature" and "culture" emerged or were reconsidered through nineteenth-century explorations of South America and the Caribbean is no accident of history. The simultaneous emergence of postcolonial concerns amidst Creole elites and subalterns who were redefining their relationship with each other and their former colonizers, and the legacy of a colonial history characterized by continued genocide and massive movements of peoples, plants, and animals after a long geophysical history of relative isolation, made it an exceptional laboratory for unsettling understandings of the world recasting the transatlantic debate on the geophilosophical (Viveiros de Castro 2011a). The place of the senses in defining the relation between the human and nonhuman is part of this long geophilosophical history.

Vision, Sound, the Colonial, the Modern

The history of the rise of the modern since the sixteenth century has been associated with the emergence of vision as the privileged sense for perception and for ideas about the subject and its relation to knowledge and the world in the West.[8] Colombian philosopher Santiago Castro-Gómez sees such ocularcentrism as crucial to the rise of "epistemic coloniality" and calls this the "hubris of the zero point" (*la hybris del punto cero*) (2004). This is the idea that "European Enlightened science presents itself as a universal discourse, independent of its spatial conditioning . . . in an imaginary according to which an observer of the social world can be placed on a neutral platform of observation [the hubris of the zero point] that, in turn, cannot be observed from any place" (2004, 18).[9] For Castro-Gómez such an emphasis on the gaze is crucial to the relation between colonialism as power and coloniality as knowledge because it gives an external observer the power to universalize its categories of knowledge and posit its own point of view as a despatialized omniscience. In Castro-Gómez's formulation, the often recognized and criticized "ocularcentrism" of Western epistemologies is not simply an accentuated tendency in Western conceptions of knowledge, as has repeatedly been cited (Jay 1993; Connor 2004; Smith 2004) but a critical element in the constitution of the colonial modern itself (Castro-Gómez 2004).

A corollary of this ocularcentric history of the moderns has been the idea that vision and sound imply opposite modes of relation to the world, what Jonathan Sterne calls "the audiovisual litany" (Sterne 2003, 15). Its main components are:

> Hearing is spherical, vision is directional; hearing immerses its subject, vision offers a perspective; sound comes to us, but vision travels to its object; hearing is concerned with interiors, vision is concerned with surfaces; hearing involves physical contact with the outside world, vi-

sion requires distance from it; hearing places us inside an event, seeing gives us a perspective on the event; hearing tends toward subjectivity, vision tends toward objectivity; hearing brings us into the living world, sight moves us toward atrophy and death; hearing is about affect, vision is about intellect; hearing is a primarily temporal sense, vision is a primarily spatial sense; hearing is a sense that immerses us in the world, vision is a sense that removes us from it. (Sterne 2003, 15)

Sterne calls it a litany because of its "theological overtones" (2003, 15) and part of such a political theology of hearing has been constituted through the idea of orality as a mode of communication opposed to writing (Sterne 2011). The identification of orality and tradition as autonomous spheres of knowledge within the epistemic domain of language has historically been a major technique for the construction of modernity and of the social inequalities within it (Bauman and Briggs 2003; Rama [1984] 1996; Ramos [1989] 2003). An aesthetic technique and disciplinary domain (orality) that contrasted with lettered elites' use of spoken and written language, and the constitution of a particular temporality (tradition as the complementary and constitutive other of modernity) became a modern paradigm for the identification of alterity within modernity itself. The creation of the field of "orality" generated a theory and methodology for lettered elites to generate a notion of alterity as constitutive of the modern. Such a field functions as a mechanism through which the subaltern is simultaneously named as having a voice, yet such a voice is subordinated by the very same principles through which it is epistemically identified as other (Martín-Barbero [1987] 2001; Bauman and Briggs 2003; Ramos [1989] 2003). This generates a complex network between culture and politics in which the value of "the people" is recognized as a political figure yet denied its political singularity, and the "other's" culture is recognized as a culture of alterity yet subordinated to the principles of high culture (Martín-Barbero [1987] 2001). Moreover since in the audiovisual complex of modernity, sound appears as the interior, immersive, and affective other of vision's prominent exteriorization, in modern formations of power such acousticity often appears as "hidden" behind the visual. In the relation between the "political theology" of orality (Sterne 2011) and the spectrality of the acoustic, the moderns generated a mechanism of power associated with the production of their own notion of alterity. Part of what is explored in this book is how such a relation between alterity, orality, and sound was constituted through the relation between the colonial and the modern in nineteenth-century Colombia. As such this is a critique of theories of the decolonial that seek to constitute "the other" by appealing to the very same mechanisms of constitution of alterity that the moderns generated.

In chapter 2 I explore how the idea of popular song in nineteenth-century Colombia was constituted as part of literary knowledge in the historical process of constituting "orality" as tradition and through different politics of writing about song. I do so by contrasting three modes in which "popular poetry," as it was called in the period, made its appearance in literature. First I look at what is understood by popular poetry and the role it was given in what is considered the first history of literature in Latin America and the Caribbean, José María Vergara y Vergara's *Historia de la literatura en Nueva Granada desde la conquista hasta la independencia (1538–1820)* (History of Literature in New Granada from Conquest to Independence, 1538–1820) first published in Bogotá in 1867. Vergara y Vergara (1831–1872) was a founding member of the Colombian Academy of Letters and a key figure of El Mosaico, an important intellectual group centered in Bogotá that functioned between 1858 and 1872 and that sought, among other things, to recognize local customs and expressions through their study and publication (Von der Walde 2007). In his literary history he gave a central role to "popular poetry" tracing its lineage to the Spanish conquest, differentiating song types according to different racial ascriptions, and positing the relation between identity, popular song, geographical regions, and climate as crucial to the nation. Vergara y Vergara was a central figure in developing the ideas that led to the transformation of blood purity as a central element of racial differentiation in the colonial period, to popular song as an element of cultural/racial differentiation in the national period. Through his work we can see the constitution of the relation between the notion of a person as having an identity and culture as a racialized category. This happens through detailed classification of the uses of the voice into distinct popular song types, "character" types, and styles. Vergara y Vergara's work in that sense is paradigmatic of the type of thinking that gave rise to the political theology of orality.

Candelario Obeso (1849–1884) and Jorge Isaacs (1837–1895) were the two Colombian intellectuals who compiled songbooks in the nineteenth century. Obeso, an Afrodescendant poet from the Caribbean town of Mompox, had a poetry inflected by detailed attention to the transcription of the acoustic dimensions of Caribbean speech, to the point that often the words in his poems are difficult to decipher at first reading. His nonstandard orthography appears as a technique of transcription that pays special attention to the acoustic in a poetry that heavily critiques the lettered city and that highlights the limits of creolization. Rather than reduce the relation between poetics and Caribbean cultural particularities to a problem of identity, Obeso highlights the incommensurability between the politics of acoustic inscription and his own audile techniques, acutely attuned to sonorous and ontological inflections that did not match the imperatives of the lettered city's progressive standardization. His work is emblematic of a process of inscription that challenged the ideas of the lettered city and that heavily auralized the format of alphabetic writing.

Isaacs, of Jewish descent and from the Andean-Pacific region of El Cauca, is one of the most cited authors in twentieth-century Colombian folkloristics due to his musically imbued fictional work, *María*. In chapter 2 I analyze his early ethnographically inflected fictional use of popular song practices as positioned between his early rise to fame as a recognized author by intellectuals in Bogotá and other parts of Latin America and his silenced heritage as a Catholic convert of Jewish descent. His early work is problematically positioned between the silencing of his own heritage and the recognition of Afrodescendant auralities, generating a complex relation between recognition and negation (Avelar 2004) that in his later work evolves into an outright confrontation with the conservative lettered politics of late nineteenth-century elites in Colombia. As we see, each of these intellectual figures is differently positioned by history and by political choice to the relation between the acoustic and its inscription, and as such they refunctionalize the ear into literary history in different ways. What one sees is that not all lettered elites had the same historical relationship to the ear as that implied by the audiovisual history of the moderns. Rather, what emerges is a more diverse and contested history of the senses in the relation between listening, vision, orality, and the politics of inscription of sound than implied by the notion of the lettered city.

One of the main questions generated by the critique of modernity in the history of the senses has been if, after all, such a history has been overwhelmingly ocularcentric or if it has been understood as such because of privileging specific practices of modernization that have made it seem so.[10] And yet, one of the most baffling issues about the "audiovisual litany" and its complex relation to the political theology of orality and to alterity is its capacity to return as an obvious construction despite repeated historical deconstructions. What Sterne calls "the political theology" of the audiovisual litany and of orality is what here I am calling the "spectral politics" (Ludueña 2010) of modern aurality—this capacity to present itself as "an other" when it is in fact "the same" as a recurrent history of the oral/aural. Even when we change the terms of reference: folklore for intangible heritage, orality for voice, racialized bodies for knowledge of the body, monotonous music for savvy rhythms, and so forth, frequently, without suspecting it, what we are doing is reproducing the same sensorial/expressive scheme that we are critiquing.

In the case of trying to rethink the relation between the colonial and the modern in Latin America and the Caribbean, we see that throughout the twentieth century, the region is often presented as having a different modernity, one that highlights the oral/aural bodily knowledge as a particular knowledge of the subaltern opposed to the ocularcentrism of the elite.[11] Such "fonocentrism" (Ramos 2010, 30) tends to take two forms: a celebration of the acoustic that limits the expression of sonic difference to the body and the voice, and a difficulty of recognizing a dense history of the sonorous and audiovisual as a field that has generated multiple modes of action, thought, and critical theorization, except when it is posited as a contrasting othering. Thus, in the name of recognizing the knowledge of "the other" such "fonocentrism" ends up reproducing an unexpected Cartesian dichotomy of the body and the mind, divided between subalterns and elites. In the name of recognizing the other, it ends up historically using the same method the moderns created to incorporate alterity into its guise, and in the name of decolonizing, it actually recolonizes. One of the main means of accomplishing this spectral alterity of the modern is through the immunization of the voice through specific vocal technologies that entangle different ideas of the people.

Chapter 4 is about the use of eloquence, etymology, and orthography as pedagogies of the voice aimed at producing an idea of orality and of music that created a notion of personhood valid for the nation-state. Following Fabián Ludueña, I call such vocal techniques used in the service of distinguishing the human from the nonhuman in the voice, anthropotechnologies, that is, "the techniques through which the communities of the human species and the individuals who compose them act upon their own animal nature with the purpose of guiding, expanding, modifying or domesticating the biological substrate with the intention of producing that which, first philosophy, and later the biological and human sciences, tend to denominate as 'man'" (Ludueña 2010, 11).

In Hispanic America and the Spanish Caribbean the political problem generated by linguistic diversity in the midst of homogenizing the nation was not only that of establishing Spanish as a proper language for both the nation-state and for Christianity (Ramos [1989] 2003; Lomnitz 2001; Rodríguez-García 2010) but also that of generating a proper mode of *voicing* it and dealing with its many pronunciations and varied folklore in an area that was no longer politically unified by the colonial dominance of the Iberian peninsula.

One of the philosophical problems of the voice in the West since Aristotle has been how it manifests the animal dimensions of the human, thus demanding a politics of differentiation of the human and nonhuman elements of the voice in the constitution of the political history of the person. In chapter 4 I explore how eloquence was used as a means to correct the fallibility of the ear and guarantee a proper relation between voicing, pronunciation, and orthography in order to produce a desired political idea of the person. Eloquence, wrote Miguel Antonio Caro, "is the art of producing sounds, words and clauses with precision and propriety, with the adequate modulation and expression, when we speak or read. Voice is the instrument of elocution and language is the essential form in which it is exercised" ([1881] 1980, 446). He modeled many of his theories on eloquence on Venezuelan philologist Andrés Bello (1781–1865), who understood the grammar of a language as "the art of *speaking* it correctly, that is, according to good use, exemplified by educated people" (emphasis mine, Bello edited by Cuervo 1905, 1). Eloquence involved then a grammaticalization of the voice in order to create the distinction between a proper and an improper human, a way of "directing the human animal in its becoming man" (Ludueña 2010, 13). Such theories were also crucial in generating the idea of culture as something that needs to be taught to the people as well as a notion of "a people" for the political processes of republicanism. Through such training the voice was understood also as manifesting an enlightened sensorial disposition of the ear that sought to curtail the fallibility of its affective dimensions.

Rufino José Cuervo (1844–1911), a Colombian philologist and colleague of Caro who wrote a large part of his work in Paris, became one of the most important etymologists of Latin America through his work on the history of words for the creation of his *Diccionario de construcción y régimen de la lengua castellana* (Dictionary of construction and regimentation of the Castilian language). For him as for Andrés Bello working in Chile, language was "a living body" (*un cuerpo viviente*) (Bello 1905, viii) and, like a living body, it was characterized by different "life epochs" (Cuervo [1914] 1987, 23). Etymological techniques emerged as the means to control language's tendency, as a "living body," toward diversification across time, by selectively determining the correct origin of a word in order to authorize its proper use in the present. The dangers of language change through inappropriate mixtures were often metaphorically expressed as akin to the dangers of promiscuous sexuality reflected in inadequate race mixtures. The most important archaisms were present in popular lore. But the history of words had to be selectively researched in order to find the most appropriate heritage in considering the value of such popular spoken and sung forms. By controlling the heritage of words and popular expressions through the proper selection of their genealogy and origin, etymology became a technique for a eugenesis of the tongue, attached to a project of national sovereignty.

As a science that combined a politics of descent with one of archaism, etymology turned time into the primary link between language, different elements of popular expressive verbal culture (later to be called folklore) and literature. Through etymology the popular was given significance as a "politics of the prior." In the governance of the prior "the sociological figure of the indigenous (first or prior) person is necessary to produce the modern Western form of nation-state sovereignty even as it continually undermines this same form" (Povinelli 2011, 15). In Latin America and the Caribbean, with the history of biological mixture, the prior had to be selectively determined. It involved not only addressing the place of indigenous languages in the nation but also the politics of highlighting Hispanic heritage while neglecting others in the description of the popular. Thus while "the people" provided the proper archaic etymological word model, they had to be trained as "a people" into linguistic propriety through eloquence. If eloquence turned the multiple into one form of speech (and one people), etymology provided the means to arrive at the definition of what or who that one should be through a careful process of vocal selection. Theories of cultural patrimony emerged through the patriarchal etymological control of language diversity by rewriting the history of the legacy of Spanish language, by producing differential theories of time for the popular and the erudite, and by developing an idea of cultural heritage that mapped a eugenesis of the body onto a eugenesis of the tongue and of folklore.

The third anthropotechnology explored in the fourth chapter is orthography, especially in its use as music notation. Composer and poet Diego Fallón (1834–1904) developed an orthographic musical notation as appropriate to the technological conditions of Colombian typography. According to Fallón, unlike the literary, which was mediated by the newspaper as a medium and the chronicle as a genre, music did not have the same means of dissemination. In his *Nuevo sistema de escritura musical* (New system of musical notation) (1869) and *Arte de leer, escribir y dictar música, Sistema Alfabético* (Art of reading, writing, and dictating music, Alphabetic System) (1885), he developed a system of musical notation based on alphabetic writing. His books show the extent to which alphabetic writing was acoustically understood as a mediation between sound and writing in this period in Colombia. He translated every single aspect of musical sound into orthographic notation in order to propose an idea of the musical work, proper for the moral edification of the citizen who was to be taught through his system. Such a system involved not only notation but a ventriloquization of musical pieces into syllabic form because prior to playing them, the student would first learn to voice the syllables that were the result of the orthographic transcription of music. Contrary to the development in Europe where the emergence of the work concept in music implied its emancipation from language (Goehr 1992), here the emergence of the work concept implied recasting the conservative relationship between music and language as a mediatic, modern one. If the literary was what gave distinction through correct speech, it was music, "the most powerful element of sociability" (González Lineros [1877] 1885, 5), an art understood as able to awaken "sweet, generous and compassionate affects in the human heart," (5) that would be the standard bearer for ethical training in the nineteenth century.

Through the relationship between theory and the political power of grammarian presidents who were invested in them, these three anthropotechnologies generated an "immunization" of the voice in the name of the formation of the political community of "the people" and in the name of an aesthetics of a proper mode of the voice. An "immunitary paradigm" is one that protects or inoculates the person through the use of the very same materials from which it wants to protect them but in some attenuated form (Esposito [1998] 2009). Protection against something through the use of the elements that cause the threat is a basic mimetic principle in sorcery and magic (Napier 2003; Taussig 1993). But in the history of Western zoopolitics of the modern, such a principle of protection becomes also one of alienation through a politics of purification that seeks to recognize something while denying the multiplicity and singularity of its constitution. Roberto Esposito associates such a paradigm specifically with the constitution of Western modernity and the notion of community. Even though for him language politics are central to the creation of an immunitary biopolitics in the name of community, he never associates that to the idea of orality.[12] In chapter 4 I explore how the aforementioned anthropotechnologies of the voice produce a politics of immunization that generates the notion of orality that became central to the political theology of the state. Orality in this book is not the opposite of writing nor a complementary "other" of aurality, as implied in the pairing of the two terms through a backslash (orality/aurality). Rather it is a historical mode of audibility that emerges in divesting the voice of unwanted features while pretending to be speaking about it. But this is not the only form of voicing that one finds in the nineteenth-century archive. The question that emerges for a decolonial history of the voice is not only how to identify the constitution of an alterity of the acoustic as part of the political theology of the state but also the presence of different modes of relating alterity and the voice that do not fit such a paradigm.

Comparativism, Transduction, and Acoustic Assemblages

As Jonathan Sterne has said, "at its core, the phenomenon of sound and the history of sound rest at the in-between point of culture and nature" (2003, 10). As we know, the main method of labor in the humanities throughout the late twentieth century has been that of "denaturalizing" what has been culturally constituted (Avelar 2013). But in "denaturalizing" the cultural constructions through which the knowledge of the "other" has been subordinated in order to recognize and reveal an "other" knowledge we often leave intact the underlying relation between nature and culture that such a knowledge implies. Instead of denaturalizing we often reculturalize by proposing new modes of representation. But not all cultures and not all peoples in different historical moments of Western history consider "nature" as the given and "culture" as the made. What is needed in altering our ways of relating the given, the made, and the sensorial is not just unsettling the history of representations but approaching the underlying relation between nature (as the given) and culture (as the made) implicit in the distinction between music and sound. That means, on the one hand, "discarding the transnational constructions about sound and the ear as a basis for the history of sound" (Sterne 2003, 10) as well as for the history of alterity in the relation between the colonial and the modern (Howes 2004; Taussig 1993). But on the other, it also means being attentive to how both nature and culture, ontologies and epistemologies, ideas about entities that listen and about entities that produce sounds are intertwined in theories about the acoustic whether understood as music, language, narrative, sound, or otherwise. This involves specifically addressing how the sonic was simultaneously constituted as a dimension of knowledge, that is, as something that needs to be judged as representations (Maniglier 2010) and as a dimension of sentience, that is, as a phenomenon that involves "an internal variability" (Maniglier 2010, 25).

According to Jairo Moreno the "cognitive gestures" that address the discursive organization of musical objects in Western music theories, in different historical periods, "invoke a figure who hears, listens and understands, as well as a means to represent what that figure hears, listens and understands" as a major strategy for addressing "the question of what and how [music] theories know what they claim to know" (Moreno 2004, 1). For him, "the cognitive allocations that condition various constructions of hearing, listening, perceiving, and understanding music by and for various subjects" (1) are crucial to the construction of theoretical writing around music. At the turn of the nineteenth century he hears the emergence "of temporality as a new domain within which knowledge of music takes place. Within this new domain there is a turn toward the existential, as listening experience and interiority become empirical addresses of musical thought" (19).[13] For Moreno then, the act of construction of theory around a particular musical object invokes a listening subject that is differentially constituted in distinct periods of Western music history. This also implies different conceptions of the ear. As noted by Ben Steege with regard to the study of acoustics in the nineteenth century: "As in many discourses about acoustical and musical phenomena, we often do not know precisely whether the 'ear' we are describing is a physical, mechanical, organic, physiological, psychological, or cognitive sort of thing. Indeed, the multiplicity of the ear's *potential* qualities and functions gives the lie to any singular notion of 'the' ear" (2012, 50–51).

Thus entities that listen and entities that produce sounds are entangled in the relation between nature and culture and mutually produce each other — a theory of sound implies a listener, which in turn imagines a listener and an idea of reception of sound. In the relation between each of these entities — a listening subject, an object that produces a sound, and a supposed listener of that sound object — what is produced is an ontology of *relationships*, an idea of how to think the interaction between entities that produce/hear sounds. That is why frequently hearing is a method that gives us the keys for how to think different ontologies of the human and the nonhuman.[14] But, in the West this tends to be confused with thinking that sound (or music) is an eminent field of transparent affect and relationality. That is why in the West the expression "to have a voice," to "listen to one another," and more recently, to feel a "resonance" or "vibrations" between people are often expressions used to invoke the idea of participation, the recognition of the "other," and alternative forms of the collective. The point is not to negate that the ear produces an ontology of the relation between the person and the world, but rather not to confuse that with our own notion of relationality. What this implies is the need to explore the richness of a multiplicity of variables among what different peoples consider the given and what they consider the made that come together in the acoustic.

This multiplicity of variables of relation between the given and the made is generated through sound/listening in what I am calling acoustic assemblages. By acoustic assemblages I mean the mutually constitutive and transformative relation between the given and the made that is generated in the interrelationship between a listening entity that theorizes about the process of hearing producing notions of the listening entity or entities that hear, notions of the sonorous producing entities, and notions of the type of relationship between them. Such an assemblage circulates between different listening entities through different practices of inscription of sound: rituals, writing, acoustic events, and so forth that, in turn, are also heard. These assemblages then imply a mutually constitutive transduction (in two directions, let us say) of notions of sound as well as notions of who listens, as well as potentially transformative processes of inscription of sound that interrelate listenings and sounding "objects." If such an interrelationship between listeners and sound objects is intercultural, that is, it occurs between beings considered "different" as is the case in colonial contexts, then we have a cycle of transductions in which each of the listening entities of this assemblage generates its own process of transformation of the relation between the notion of the listening entity, the notion of the sound producing entity, the process of (re)inscription of such hearing and the type of relation constituted in the process.

In such an assemblage we have less a transparent field of acoustic communication as implied by the audiovisual litany than ample possibilities for equivocation. We can link this to the idea of transduction associated with the study of the senses. The notion of transduction means the transformation of one form of matter or energy (sound waves, light) into another (vibrations, biochemical transmitters, etc.).[15] Anthropologists have also thought about the intercultural context of mutual encounters of alterities through the notion of transduction (Viveiros de Castro 2004; Helmreich 2007).

Eduardo Viveiros de Castro has proposed a renewal of anthropological thinking from a "method of controlled equivocation" (2004, 3), derived from Amerindian thinking, in which it is assumed that there are different perspectives in the conceptualization of an entity. As such the modes of conceptualizing the relation of difference do not depend on the history of social constructivism (Holbraad 2012). The work of a comparative anthropology would not be to translate the concepts of the other into its Western equivalents in order to adequately "explain" who the other is, to represent him or her divested of colonial history—a process that assumes that I, as decoder, am able to carry the weight of truth about a "correct" reading of the other; it would rather involve assuming differential fields of conceptualization, each with their potential equivocations as a field of comparative mutual constitution of notions of alterity.[16] Equivocation is not a mistake. To the contrary:

> Anthropology then is interested in equivocation, in the literal sense of *inter esse*, being in between, existing in the middle ... The crucial point here

is not the empirical fact of incomprehensions but the "transcendental fact" that they are not the same. The question then does not consist in knowing who wrong and much less in knowing who cheats whom. Equivocation is not a mistake, nor a confusion, nor a falsity, but the very basis of the relation that implies it, which is always a relation of exteriority. . . . Equivocation in sum is not a subjective failure, but a dispositive for objectification. It is not an error nor an illusion—it is not about imagining objectification in the language of reification or fetishization—but the limit condition of every social relation, a condition that is itself overobjectified in the case of the relation we call "intercultural," in which the language games diverge to the extreme. (Viveiros de Castro 2010, 78–79).

For Viveiros de Castro such a process of mutual equivocation for a comparative anthropology implies "not a process of induction, nor deduction, but of transduction" (2004, 20). Following Simondon, he takes the idea of transduction as a model for the notion of controlled equivocation in a decolonial anthropology. Here it is assumed that "difference is a condition of signification and not an impediment" (20). Stefan Helmreich in his study of transduction of sound underwater also proposes the idea of "ethnography as transduction" (2007). He critiques notions of both sound and ethnography as spheres that give access to truth through a process of immersion in their matter. Rather, for him, "a transductive ethnography would be a mode of attention that asks how definitions of subjects, objects and field emerge in material relations that cannot be modeled in advance. Most modestly I offer it as one idiom for thinking through anthropologies of sound . . . More expansively I suggest that a transductive ear can help to audit the boundaries, to listen for how subjects, objects, and presences—at various scales—are made" (2007, 632).

The notions of transduction of Viveiros de Castro and Helmreich are similar in that both suspend the idea that establishing relations in difference means arriving at "the truth" of the other in order to explain it and question rather how "subjects, objects and presences" mutually constitute each other. The idea of this transductive anthropology along with the questions raised by the recent histories of the senses place us at a historical moment when we question the idea of whether the supposed ocularcentric history was actual or if its hypervalorization actually happened because of historiographical practices that do not recognize the importance of other senses and other histories of the senses. The question that emerges is either whether these alternative sensorial histories have simply been there as "subterranean forms of auditory knowledge" (Hirschkind

2006, 121) that were at the margins of a mainstream dominated by the audio-visual complex, or whether a recasting of such histories gives rise, in effect, to a different temporal and hierarchical cartography of the senses (Hirschkind 2006; Howes 2004). Taking into account non-Western histories of the sensorial is a central way of rethinking the ways in which anthropological accounts of the senses are significant not only for a particular place but for a global history of the senses. One of the elements that emerges when doing so is that the relation between the inscription of the acoustic in the musical, the folkloric, and the linguistic, on the one hand, and in sensorial and broader life histories, on the other, is the multiple temporalities (biological and cultural, geophysical and social, economic and material) that accrue in the fact that listening is simultaneously a physiological, a sensorial, and an interpretive cultural practice. Taking into account such temporalities opens up our understanding of the global histories of the colonial/modern not just to a geopolitics of knowledge and economy but also to its relation with the geophysical and ontological. This takes us back to the relation between the history of comparativism in the nineteenth century and its place in the formation of histories of ideas in the twentieth century, especially regarding the place of the voice in the identification of language, song, and music as spheres of knowledge.

The history of nineteenth-century comparativism in Latin America and the Caribbean followed different trajectories from those of Europe, even though both were mutually constituted in the exchange of ideas between them. In some cases it was used to rearticulate new forms of exclusion through a racialized culturalism that used the comparative method to transform the politics of blood purity into cultural theories of discrimination (chapters 2 and 4). But it also gave rise to a relation between the abundance of nature and knowledge that became particularly significant at the end of the nineteenth century and in the first half of the twentieth century, in that it began to formulate the anthropological as a site of expressive creativity that would come to be recognized as unique and highly significant for the history of music, language, literature, and their respective disciplinary histories in the region. In particular, this relation between the magnificence of nature and the joy of knowledge it produced was a way of handling the limits and incommensurabilities of perceived processes of equivocation in such spheres as language, song, music, and forms of narration. In Colombia, listening to indigenous languages played a crucial role in recasting the juridical place of indigenous groups in the nation through the political theology of the nation-state (see chapter 4). But it also generated a type of *naturphilosophie* that centered on the recognition of the relation between indigenous mythical histories, the seeming abundance of tropical nature, linguistic particularity in order to, at least, raise the question of a different history of "man." Such questions were related to an emergent recognition of the value of indigenous languages and vocal expressive practices—myths, songs—contra their negation or useful only for purposes of conversion.

In this book, through the history of comparativism in the region, I question the conceptual, temporal, and spatial framing of this history by rendering it not solely as a European one but one produced in the global trade of ideas about expressive culture and the type of making and doing we call art in the relation between the colonial and the modern. As stated by Ticio Escobar, in Amerindian history, such a "tissue of sensible experience" appears deeply entangled in particular networks of collective experience, functions, and rituals, that "reveal, in the play of form, dense truths that are inaccessible through other means" (Escobar [1993] 2012, 32). For the purposes of this book then, one can think of such a history in the Americas as presenting contested understandings and uses of "ways of making" with sound that entangle the ontological and epistemological in the manipulation of design (Ingold 2011). Some of the aspects of such a history highlight how the recognition of indigenous languages, modes of narration, and different vocalizations, by nineteenth-century intellectuals, raised contested political understandings of indigenous groups and their modes of narration in the new nation. Particularly salient was how incommensurate modes of interaction gave rise to new genres, such as colonial grammars, letters to colonial authorities, and musical rituals that emerged from the colonial process (Hanks 2010). The history of comparativism helps us understand how the politics of equivocation regarding particular expressive practices led to a politics of expressive transformation as a mode of political response. This sometimes generated a conflictive zone of recognition that questioned the very relation between governmentality and expressive indigenous practices proposed by the nation-state.

In nineteenth-century Colombia, two Colombian scholars, Ezequiel Uricoechea (1834–1880) and Jorge Isaacs used and adapted comparative methods and ideas in their studies of indigenous languages in the transition toward the formulation of ethnography and linguistics as disciplines, and of indigenous expressive practices as a significant aspect of the literary and of expressive culture in general. They did so as humanists invested in the development of the appreciation of the local, in order to answer questions about the nature and history of the American continent. In chapter 3 I explore the theoretical significance of thinking the relationship between the valorization of indigenous cultures and the political in the dialectic between nature and culture that emerges in the fracture between hearing and writing indigenous languages.

Uricoechea and Isaacs stand out as unique and exceptional figures in their approach to indigenous languages, after centuries of their study for purposes of religious conversion and in the midst of the rise of nationalistic language policies that sought to eradicate them.

Uricoechea was a Colombian naturalist and philologist who spent his life between Colombia, the United States, and several European countries. A scholar who self-defined himself as passionate "for all things American," he founded the *Collection Linguistique Américaine*, in which he sought to critically edit the indigenous grammars collected by missionaries during the colonial period in Latin America and the Caribbean. He was also one of the early archaeologists in Colombia, and founder of the *Sociedad de Naturalistas Neogranadinos* (Society of New Granadian Naturalists). I explore how his work with indigenous languages was related to the question of the nature of the American continent and its geophilosophical significance for the emergence of indigenous linguistics and for an aesthesis of the local.

After the publication of *María* in his early twenties, Isaacs became a soldier who fought in the ranks of radical liberals against the conservative government but who never lost his passion for local popular expressive forms. Toward the end of his life he wrote one of the earliest ethnographies in Colombia, *Estudio sobre las tribus indígenas del Magdalena* (A study on the indigenous tribes of the Magdalena region) (1884), a text that generated a virulent response from philologist and politician Miguel Antonio Caro entitled *El darwinismo y las misiones* (Darwinism and the missions) ([1886] 1980). Caro was nominally vice president but actually acting president of the nation between 1892 and 1898. He was the author of the Constitution of 1898 that was to last, with several reforms, as the nation's Constitution until 1991, and a key figure in the establishment of the relationship between language, jurisprudence, and the law as a politics of the state. In the hands of Caro and other "grammarian presidents" philology became an instrument of power, not only in the general sense of a relation between power and language but, specifically, in the relation between the exercise of jurisprudence, the knowledge of philology, the politics of civil war, and the institutional use of political rhetoric by the state (Deas [1992] 2006). In this chapter I analyze the significance of the political controversy between Isaacs and Caro and its legal implications for indigenous languages and peoples in the midst of redrawing the boundary lines between nature and culture, between the sound of languages and the politics of their inscription, and between war, politics, and the law. Here the emerging tension between missionaries and ethnologists and their role in the national politics of indigeneity was central to defining the value of indigenous languages for the nation-state.

If a proper aural corpus to represent the Colombian nation has been historically seen as lacking, materials about practices of hearing the voice, as well as other sounds not included in this study, such as instrumental music, are evidently not. The particular archival material explored here in each chapter is only a fragment of what I found. When one listens to the historical archive, without looking for the genealogy of a particular musical genre, but rather simply exploring the way listening practices are found across different forms of writing, what emerges is a series of practices of listening and sounds that extend beyond our present-day ideas of what counts as a proper genre, music, or language. Listening to vocalities was used to establish the historical divide between the colony and the postcolony by defining the nature of different peoples through theories of vocal propriety for the new nation-states.

In the midst of very different political positions and ideas, the nineteenth-century intellectuals studied in this book were dealing with similar questions: if in the new nations all were to be deemed citizens and therefore had to be politically defined as persons, then what counted as a proper human voice? How was that established in the midst of a colonial history that left a legacy of discourses and practices about the questionable validity of the natural history of the continent? How was the juridical status of humans who had historically been considered and treated as not belonging to the juridically valid political community of persons to be redefined? What about the (mis)hearings generated for indigenous or Afrodescendant peoples for whom "becoming animal" or other forms of nonhuman becoming through the voice was precisely one of its many powerful uses? What were the implications for them of assuming a politics of the voice as representative of an autonomous individual in the politics of the nation-state? How did descendants of Hispanics as well as Afrodescendants and mixed peoples make sense of their belonging in a continent to which they could not trace their original heritage and of a heritage that was territorially dispersed? How was indigeneity, and the practice and study of indigenous expressive culture redefined through the politics of nationalism? These questions were not unique to the groups of people or intellectuals described here. Rather they were common to other countries in Latin America and the Caribbean because the political moment defined by the postcolonial struggle generated similar questions even though they were answered differently in each place. Thus while the questions explored here involve addressing the form the answers took by virtue of a particular archive in a particular place, many of these issues were also being considered and explored in other parts of Latin America and the Caribbean. One can think of this book as a microcosm of a broader history of the relation between the ear and the voice, one that implies a shared history of cosmopolitics (that is, a politics that implies taking into account both humans and nonhumans, Stengers *dixit*) for the region even when the particular histories differ. The study of the relation between the history, the voice, and the lettered city is thus a geophilosophical problem, not just an epistemological one.

The contested nature of the person and of nature have recently become major critical concerns, partially as a response to the unprecedented devastation of the environment produced during the past 250 years or so. Latin America and the Caribbean enter the present critical juncture, not only through their centrality in articulating the critical role of present-day environmental struggles through indigenous movements and through the political struggles around crucial regions for global environmental politics like the Amazon. The region also contributes a long and contested history of the person and of nature and therefore brings to the foreground alternate histories for narrating and understanding such a crisis. The relation between the ear and the voice explored in this book is part of this broader history.

1 • ON HOWLS AND PITCHES

On April 19, 1801, Alexander von Humboldt began his trip up the Magdalena River en route to Bogotá, then the capital of the Viceroyalty of New Granada.[1] During this trip he undertook experiments with the horrid smelling vapors expelled from the mouths of crocodiles, wrote essays comparing the mosquitoes from the Orinoco, Rionegro, and Magdalena Rivers, contrasted the exuberant vegetation of the Magdalena riverside to the orderly growth found along the Rhine, and wrote observations on the more than eighty *champanes* (boats) that one could see transporting contraband to the beautiful riverine port city of Mompox and other cities of the Colombian Caribbean during times of peace. As if re-creating the seven days of Genesis, Humboldt remapped the lands, flora, and fauna of the Magdalena River onto the scientific observations of his inner cosmology. These episodes barely made it into his famous *Personal Narrative of Travels to the Equinoctial Regions of America, 1791–1804, by Alexander von Humboldt and Aimé Bonpland*, a public account of his travels in the Americas that stops short of his journey from the Colombian Caribbean into the Andes. Thus many of his thoughts and observations on Colombia recorded in his unedited diary, remained unknown until well into the twentieth century (Arias de Greiff 1969).

Perhaps due to its more intimate nature, the diary gives us a glimpse of those things that unnerved Humboldt. And try as he might, there was something that he found impossible to contain through scientific observation: the sound of the *bogas*, the boat rowers of the Magdalena River. His positive impression of their tremendous physiques and "demonstration of human force" (which he would have liked "to have had to admire for less time") was muted by the sounds they made:

They are free men, sometimes very arrogant, unruly and happy. Their eternal happiness, their good nutrition . . . all of this diminishes the feeling of compassion for them. But the most upsetting thing is the barbarous, lustful, ululating and angry shouting, which is sometimes like a lament and sometimes joyful; at other times full of blasphemous expressions through which these men seek to handle their muscular effort. About this point we can make quite a few interesting psychological observations. All muscular effort decomposes more air in the lungs than during repose. To bring more air into the lungs, it is also necessary to expel more vitiated air. That is why, in heavy work, the emission of cries and sounds is quite natural. If the type of work has a regular cadence (wood cutting, rock drilling in mining, the setting of sails by sailors) then a psychological factor is added. The pleasure for cadence requires that the tones be expressed in a more determined way: Hau Hau. . . . . . Ham, Ham. . . . . Halle, Halle. . . . if you add all that you can imagine, the tone can become a song and even a dialogue. Thus, the heavier the work, the more angry the screaming of the bogas, among whom the cadence will be affected frequently by caprice. They begin with a sibilating has has has and end with exacerbated insults. Especially, each bush from the shore that they can reach with the pole is saluted in the most improper fashion, the has rapidly turns into a bellowing ruckus, into a blasphemy . . . The racket you hear uninterruptedly until you reach Santa Fé (Bogotá) is as bothersome as the steps of the bogas on the roof of the champán, over which they stomp so loudly that frequently there is a threat of it collapsing. Our dogs needed many days to get used to this unbearable racket. Their barks and howling increased the scandal. (Humboldt 1801, 29)[2]

For Humboldt, scientific observation, the means of making sense of radical difference in an intensely heterogeneous context, was drastically unsettled by an "acoustic release" (descarga) (Ramos 2010) that made it impossible for him to interrupt the racket of the bogas. In this brief passage, sonic perception is spread on corporeal difference, scientific explanation, and the narration of uncontainable, bodily produced noises—vocal utterances and stamping feet—that penetrated the ears without interruption throughout the day and over the course of several weeks of travel. The description of sound in this passage stands out from the larger corpus of Humboldt's writing because of his repeated use of negative adjectives of excess—barbarous, lustful, angry—only one of which, ululating, actually refers to sound itself. All the other adjectives used here metaphorically map themselves onto words that express a lack of emotional and bodily containment that are often associated with the irrational.

We also hear the difficulty of deciphering a generic category of sound. Was this a lament or a joyful type of expression? What about the blasphemy and the racket on the roof produced by the bogas' stamping feet? It was a sound that was impossible to inscribe onto a genre or an emotion, its untraceability begging for classification in Humboldt's ears. The ephemeral nature of sound is supposed to be one of its defining qualities, but when sonic perceptions are troubling, or perceived as unwanted, then sound becomes endlessly unbearable, materialized on the body as a sign of the limits of listening as a dialogic practice. It is as if Humboldt found it difficult to overcome his acoustic disgust in order to undertake the project of epistemologically mapping sonic difference as scientific observation.

As Steven Feld reminds us, "sound, hearing, and voice mark a special bodily nexus for sensation and emotion because of their coordination of brain, nervous system, head, ear, chest, muscles, respiration, and breathing" (1996, 97).[3] Such sounds are then interpreted and experienced under what Feld has called the "local conditions of acoustic sensation, knowledge and imagination embodied in the particular sense of place" (1996, 97) which he calls an acoustemology. By this he "means an exploration of sonic sensibilities, specifically of ways in which sound is central to making sense, to knowing, to experiential truth" (Feld 1996, 97). In this case, what we are able to hear through the pages of history is the contrasting perception of those who produced the sounds and those who listened to them, as mediated by potentially radically different interpretations of the same sounds since, evidently, the bogas were not bothered by their own sounding. Once sound is described and inscribed into verbal description and into writing it becomes a discursive formation that has the potential of creating and mobilizing an acoustic regime of truths, a power-knowledge nexus in which some modes of perception, description, and inscription of sound are more valid than others in the context of unequal power relations. And yet, in these colonial contexts of intense contact, one has to wonder how the boundaries between one form of knowledge and another interact, even if it is in a context of unequal power relations.

Knowledge in sound often confounds the boundaries between sensorial perception and discourse, between nature and culture characterized by sound's capacity to reverberate in the body and in different entities. Claude Lévi-Strauss drew attention to this characteristic, stating that "in music the mediation between nature and culture that occurs within every language becomes a hypermediation" ([1964] 1983, 27). Augoyard and Torgue (2005) explore what they call the "sonic effect," which calls attention to "the relation between the observer and the emitting object" (8) that is formed between "the characteristics of the constructed environment and the physical conditions of hearing and listening" (9). Although I do not think of sound as an "effect" but rather as an event whose emergence simultaneously transduces bodies and multiple entities, I wish to use their linking of circumstances, experience, and vibration to incorporate the idea that an acoustemology is forged not only by "the ways in which sound is central to making sense" but also by the ways in which acoustic knowledge is located at the nexus of what we are able to make sense of and what is beyond sense making but still affects us. As such, the experience of such knowledge is not only articulated by how human beings make sense of the acoustic through words but also by the very allure of the acoustic, by the relation between the capacities of sound to affect different entities and of different entities to be affected by sound. In the experience of acoustic perception in contexts of social heterogeneity, emotional and discursive knowledge of self and other, perceptual and descriptive knowledge of sound, and descriptions of the allure of the sonic are often collapsed into one another.[4]

Connor has stated that "perhaps the most distinguishing feature of auditory experience is its capacity to disintegrate and reconfigure space" (2004, 56). Studies of acoustic perception show that "because sound, itself, has no spatial properties, sound localization itself is based on perceptual processing of the sound produced by a vibrating object" (Yost 2001, 440). The forced proximity of the bogas and the passengers in the champán collapsed auditory regimes best kept at a distance, placing them in an unsilenceable world whose temporary spatial reordering was caused by the riverine transportation technology that prevailed in Colombia during this period. In this chapter I wish to explore the ways in which the perception of acoustic difference—that of the bogas, that of an exuberant natural world, and that of the riverine population along the Magdalena—was made sense of and mapped onto the practices of acoustic knowledge-making by Europeans, Creole elites, and the bogas.

Here listening is understood as "a historical relation of exchange" (Novak 2008, 16). The recognition of the role of listening in the constitution of acoustic ontologies and knowledges complicates the idea of how notions of "local" sounds or musics emerged, and questions the epistemological construction of local sounds as static traits meant to represent a particular place and people. Rather, "the emergence of new musical genres [or the materialization of sonic perceptions across acoustemological differences] is an on-going cycle of multi-sited, multi-temporal interpretations which must be situated within a global history of exchange" (Novak 2008, 16). Moreover, a description of music does not necessarily conform to its practice (Perlman 2004) or constrain its capacity to affect different persons, even ones belonging to the same group, in radically different ways. So, this is not a history about clashes between different musical "traditions." It is, rather, a history of how different notions of convention and invention (Wagner [1975] 1981), of what is given and what is made, coalesce in discussions about the nature of sounds and music, and of the entities that produce them in a world that had undergone drastic upheaval due to colonization.

Because of their role in connecting regions and peoples, transportation technologies in this period acted as communication technologies, conflating spatial and communicative regimes. In the colonial period in Colombia, boats were manned by bogas, or boat rowers, identified mainly as *zambos* by the eighteenth century, men of mixed Amerindian and African origin, who by then held a virtual monopoly on river transport and as such became central characters of the many types of passages initiated by travel. During the colonial period, rivers provided the main form of transportation and communication in Colombia, a country repeatedly characterized, since the conquest, as having a difficult geography, fragmented not only by the division of the Andes into three distinct mountain ranges but also by the dense tropical rainforest vegetation of the lowlands. The Magdalena River became the leading navigational route in terms of sociopolitical importance for the formation of the nation-state because it connected the Atlantic with Bogotá, the capital of the new republic, and with Antioquia, a primary gold mining state. It was also the route of entry into the Andes if one wished to go from the Caribbean into the larger South American Andean region by land instead of by sea.

Before the introduction of the steamboat in Colombia in the mid-nineteenth century, and even afterward (since steamboats were used only in parts of certain routes), travel with bogas on the comparatively small champanes was the only means of transportation in the Magdalena River, the main route of entry into the country. The champanes were large dugout canoes that were covered in the middle by rounded, thatched roofs, a design feature that was supposedly imported from Asia in the sixteenth century (García Bernal 2007). (See figure 1.1.) The bogas stood on top of these roofs, alternately pushing against or raising the long poles that they pressed against the bottom of the river to make the boats move.

Depending on its size, each champán was manned by a crew of seven to eighteen bogas.[5] According to the weather, travel between the Caribbean and Bogotá took between six weeks and three months, and was conducted in two stages: first from the ports of departure on the Magdalena River's outlets in

FIGURE 1.1. • Bogas from the Magdalena River by Ramón Torres Méndez (1809–1885). Champán en el Río Magdalena, Colombia, 1878. Litografía en color (Lithographic Ink/ Industrially manufactured paper). 25.5 x 34.7 cm. Reg. 3776. Colección Museo Nacional de Colombia. Foto: © Museo Nacional de Colombia/Ángela Gómez Cely.

the Caribbean to Mompox, and then from Mompox to Honda, if the destination was Bogotá, or to Nare, if the destination was Antioquia. Humboldt's trip took forty-five days (Humboldt 1980). One day of navigation toward Bogotá, upstream counter current for ten hours on a heavy champán, covered fifteen kilometers of navigable terrain. The same vessel could cover thirty to forty kilometers a day when traveling downstream, north toward the Caribbean. The champán thus became the site of prolonged encounters between different types of people. Acoustic exchanges acquired a particular density due to the great amount of time travelers and bogas spent in close proximity. Humboldt was not alone in committing his acoustic impressions to writing, and the howling of the bogas is a recurrent topic in travel writings of the period.

The nineteenth century was a crucial period for the constitution of the disciplinary formations that, in large measure, still persist until the present. The construction of natural sciences was itself mediated in good measure by travel literature of the eighteenth and nineteenth centuries,[6] with the natural sciences in turn serving as a model for the disciplinization of musicology and comparative musicology at the end of the nineteenth century (Clark and Rehding 2001). Among others, Jorge Cañizares-Esguerra has shown how natural history, as cultivated in the Ibero-American world during the eighteenth and nineteenth centuries, was crucial to the construction of the epistemic transformations of modernity, as important as the math and physics of Northern Europe, which are a more frequently recognized site of scientific consolidation (2006). If today the idea of "nature" has come into question, then so does the history that has articulated its different definitions. In this chapter I explore different practices of listening to voices, their role in the construction of different notions of human nature, of "nature," music, and sound. Specifically I explore how natives' howls seemed to be the limit against which Western ideas of music took form in the late nineteenth century. I begin by comparing how different Creole or European travelers perceived the vocalizations of the bogas.

A Cartography of Sonicities

The voice of the bogas and the sound of the Magdalena River basin in the eighteenth and nineteenth centuries is a legacy found in snippets and fragments, an exceptional audibility that either interrupts or accompanies, sotto voce, a narrative meant to highlight what is seen in a voyage. What are the practices of interpretation through which such sounding is described and comes to be associated with particular types of personhood and to particular ideas about nature? I begin to explore this question by comparing the different travelers' testimonies of their sonic perception of the bogas. This comparison allows us to highlight some of the terms and traits that appear repeatedly across different testimonies, thus creating a historical account of a particular sound that was described, again and again, through similar acoustic interpretations.

Auguste Gosselman, a Swedish botanist who traveled through Colombia between 1825 and 1826 and who published his travel book in 1830 wrote:

> When one of them [the bogas] pushes in a certain direction, the other has to do it in the opposite sense, after which he runs from one side to the other, howling like a dog, and in the midst of screams and whistles comes back in the opposite direction to initiate the chore again. Thus, all day long, at a temperature that, in the shadows fluctuates between thirty and forty degrees [centigrade]. (Gosselman [1830] 1981, 102–3)

As with Humboldt, the sound of the bogas is heard as a function of physical labor but is mimetically imagined as the sound of howling dogs. Despite the racket of the bogas, Gosselman was also able to hear the silence of the river and the sound of animals: "During the morning the route followed the left margin of the river, with the company of monkeys and parrots as the only ones capable of interrupting the silence of the river" (Gosselman [1830] 1981, 104).

Charles Stuart Cochrane, a captain of the British navy, published his travel book *Journal of Residence and Travels in Colombia during the Years 1823 and 1824* in London in 1825 and in Jena, Germany, that same year. Cochrane traveled between Santa Marta and the Magdalena River on a small canoe through the *ciénaga* (marsh) and canals that connected the river to the city. Upon reaching the Magdalena River he saw a large champán for the first time:

> A little before we entered the Magdalena, my notice was attracted by shouts and cries which proceeded from the bogas forming the crew of a large champán, alongside of which we presently found ourselves, the channel being barely wide enough to allow us to pass; we thus had an opportunity of witnessing the ridiculous gesticulations used by these people in the practice of their toilsome vocation. They push forward the vessel by means of poles twenty feet in length, against which they lean with their breasts, uttering a sound somewhat resembling that with which an English groom gratifies himself, while rubbing down a horse—"huss, huss, huss;" diversifying this monotony with a variety of cries and ejaculations, whilst they keep up a pantomime of bodily contortions, stamping, dancing, wriggling and twisting in a thousand ludicrous postures, unutterable and inimitable which they renew, with increased zeal, and unbounded satisfaction to themselves, the moment they perceive they have fixed the attention, or excited the laughter of strangers. (Cochrane 1825, 74–75)

Further on, traveling down the Magdalena River on his hired champán, he states:

> We found the continual pastime of lying down in the canoe very tedious, and were very much annoyed by the noise of the bogas, who accompany their work with a variety of uncouth sounds, stamping also most violently at intervals on the toldo [roof] over our heads, on which four were usually stationed to work, the other two being on the part before us. Every time they stopped to make their meals, one of them uttered a prayer, invoked not only the Virgin, and all the Saints in the calendar, but many more of their own invention, praying for a prosperous voyage, and safe return to their families; they bestow also on such who have incurred their displeasures a variety of opprobrious epithets,

at the end of which they cross themselves thrice with great rapidity; first, on the forehead; secondly, on the nose and cheeks; and lastly, from the top of the head to the waist; at the same time pronouncing Amen. They thus make fun of a custom which they would, nevertheless, think it wrong to omit and which no doubt originated in piety. (1825, 142–43)

Songs to different spiritual entities and "opprobrious epithets" to those that displeased them then were common among the bogas, performed at every site in which they stopped. Isaac F. Holton, a North American who traveled in New Granada between 1852 and 1854, whose book *New Granada: Twenty Months in the Andes* was published in 1857, spent a week traveling up the Magdalena River in a champán with seven fellow passengers or, rather, "prisoners and victims with whom I was now brought into so close and involuntary an intimacy" (81). Holton and his party were at the mercy of "an uncivilized horde of bogas, most of them absolutely naked" (81), with whom they were forced to bargain in order to be able to travel.

These [the bogas] all assembled in the front open space, the forecastle; and one of them began a prayer, which all the rest finished. I could never determine whether this prayer was in Latin, Spanish or Lengua Franca. Then, most of them sprung to the roof, seized their palancas (poles), and commenced pushing against the bottom of the river, and walking toward the stern shouting Us! Us! Us! Us! Us! Us! till they could go no farther. Their cry was tremendous. Oh for some method incapable of exaggeration, like the photographic process, to record it and compel belief! A pack of hounds may make as much noise in some given half hour as a crew of bogas, but these continue it, only with the intermission of eating and crossing the river, from daybreak till night. They shout and jump on the toldo [roof] over your head till you might fancy them in battle and repelling boarders. (1857, 82)

J. J. Borda, a Colombian educator and poet, included a piece titled "Seis horas en un champán" (Six Hours in a Champán) in one of his *Cuadros de Costumbres* (1866), a literary genre of the period used to chronicle local customs and habits:

To my side a parrot with brilliant plumage was shouting non-stop. At the stern twelve semi-naked mulattoes had their paddle in their hands and four others on board would raise their thin levers to support the work of the rowers by placing the end of their levers on the trees on the riverside and the other on their chest. Suddenly we heard a general shout, an invocation to all the saints in the calendar, mixed with the most obscene words and the most vulgar exclamations. It was the good bye of the bogas: the heavy champán began its route at the mercy of the current. The twelve rowing bogas, placed half and half to one and another side of the stern, would raise and let fall the paddles in time, stamping their callused feet, screaming in excess, imitating the sound of the tiger, the whistling of the serpent, the shout of the parrots and the voice of other animals. The champán would rupture the murmuring water. . . . all of nature seemed to become more beautiful, move, smile and sing. (Borda 1866, 285)

Finally, Auguste Le Moyne, a French diplomat in Colombia between 1828 and 1839, gives a particular twist to all of the testimonies cited above.[7] He wrote his text toward the end of his life, and thus the book recounts what was by then a distant memory of travel,

I have become convinced of the truth of what many travelers had told me about the development of the senses of sight and hearing among the inhabitants of the savage regions where there exists the necessity of discerning between the multiple screams that interrupt the silence of those lonely places; those that reveal the more or less close presence of each species of animals—inoffensive or dangerous—makes them constantly ask themselves about distant sounds and hidden objects in those vast horizons. During our navigation sometimes we would see on the surface of the water a distant point that we could hardly distinguish, when the bogas were already telling us it was a boat and they could appreciate its size and count the boatmen on it, calling them by their names and recognizing those who were their friends among them. They would call them by their names and ask them about their health and about the incidents of the trip. As to the responses—which certainly they heard, because they answered—they were only perceived by our ears as confusing sounds. We were even more admired that such men could recognize each other at such distances, when, as I have already said, they were almost naked and the particular forms of their bodies and faces are hard to distinguish due to the black or bronze color of their skin. (Le Moyne [1880] 1985, 76)

From the repeated traits found in these and other testimonies, it is possible to reconstruct some of the characteristics of the sound and performance of the bogas who manned the boats on the Magdalena River. All the trips began with prayers to the Virgin Mary that were linguistically structured not only by words considered prayers by whites but also by what they considered "blasphemous epithets" sung to different deities and also against those who displeased them. Linguistically, such prayers defied any easy identification in the ears of travelers, since the presence of multiple languages was one of the characteristic traits of the polyglot Caribbean, where indigenous and African languages mixed with Spanish and the Latin of Catholic invocations. Once on their way, the vocalization used to accompany the use of the poles for moving the boat was one that blurred the boundary between speech, melody, and shout. It was sung simultaneously by several of the bogas and continued non-stop while they worked with the poles. Such simultaneity defied a presence of either a clear tone or harmony since sounding like animals was the most common comparison. Such vocalizing only took place when navigation was conducted with the poles (it was not done when a different rowing technique with oars was used, for example, to cut across the river) and was always accompanied by stomping feet. It was also what today would be called improvisatory in the sense that it incorporated acoustic references (blasphemous ones according to several travelers) to the bushes, trees, and surrounding nature found along the way. Bogas repeatedly made use of a syllable or vocable that all of the written testimonies describe in similar manner—huss, hum, halle. Functionally, these vocalizations were used to accompany labor yet all bogas are described as tremendously irreverent, unruly, and with exaggerated bodily contortions. The bogas seem to have a penchant for what we call musical hybridity or fusion, "badly imitating" fragments of prayers that turn into "blasphemy" in the ears of Europeans, a remix practice that also involved vocables and the acoustic incorporation of the sounds of natural entities around them, all in the rhythmic regularity of vocalization for labor that involved repetitive movement. According to Le Moyne bogas were also sensorially acute, capable of clearly distinguishing words and identifying peoples across great distances.

But however accurately we might be able to reconstruct the sound and performance practice of the bogas, I am more interested in what we can learn from the labor of interpreting the underlying notions of sonic creativity and use that seem to appear repeatedly in these statements than in trying to discover how colonial Colombia sounded. I am also not interested in trying to identify a local "genre," but rather in exploring the interpretive gap that transpires between the moment "bodies are affected by rhythms, frequencies and intensities before their intensity is transduced by regimes of signification and captured in the interiority of human emotions and cognition" (Goodman 2010, 132) and the moment that ensonification is accomplished through inscription into writing.

The acoustic intensity felt by European and Creole travelers in the Magdalena River was produced by multiple factors. The most obvious factor is the bogas' howling which, through acoustic resemblance, are repeatedly compared to different animals. As Gary Tomlinson has shown in his study on singing in the New World at the moment of European contact, such "heightened voices" appear repeatedly in the colonial archive yet have been resoundingly ignored by histories too concentrated on speech and the lettered word (2007). The vocalizations of the bogas and many others in the colonial archive seem to defy a description as either speech or song, and are thus likened to animal sounds. The surrounding landscape is accoustically perceived as a background "silence," that noisy humans and animals interrupt. Le Moyne also suggests that the acute audiovisual sensibility of the bogas is a predatory one needed to identify animals. This acoustic sensibility has been a trait repeatedly explored in musical ethnographies of the rainforest that emphasize the sensorial tuning to species and spatial acoustics (Samuels et al. 2010). Humboldt also ascribes to the bogas one of the theories of the origins of music prevalent in the period in Europe—that music originates in the rhythmic movements of labor. And the bogas imitate and copy acoustical fragments performed in different languages in order to bring them into their vocalizations. The elements we have then are multiple uses of imitation; a relation between physical movement, sound, and the labor of navigating; mixture of sonic fragments that come from different sources, a fact that is seemingly untroubling to the bogas; chanting to multiple human and nonhuman entities, both spiritual and animal; sensorial acuity, noise, and silence. All of these are classical acoustic themes of the colonial archive. Let us use a classic trope of interpretation from what is, by now, a classical text on the topic by Michael Taussig as an entry point into the question of exploring the type of acoustic knowledge at work here:

> The wonder of mimesis lies in the copy drawing on the character and power of the original, to the point whereby the representation may even assume that character and that power. In an older language this is "sympathetic magic," and I believe it is as necessary to the very process of knowing as it is to the construction and subsequent naturalization of identities. But if it is a faculty, it is also a history, and just as histories enter into the functioning of the mimetic faculty, so the mimetic faculty enters into those histories. No understanding of mimesis is worthwhile if it lacks the mobility to traverse this two-way street, especially perti-

nent to which is Euro-American colonialism, the felt relationship of the civilizing process to savagery, to aping. (1993 xii–xiv)

The history of the mimetic faculty invokes two ontologies and epistemologies of such a faculty, what Taussig calls "sympathetic magic," on the one hand, and representation as leading to identity, on the other. How does this mimetic faculty enter history through the howling of the bogas? Jacques Rancière associates the mimetic principle to a particular regime of the arts he calls the poetic or representative regime. For him, "the mimetic principle is not at its core a normative principle stating that art must make copies resembling their models. It is first of all a pragmatic principle that isolates, within the general domain of the arts (ways of doing and making), certain particular forms of art that produce specific entities called imitations" (Rancière 2006, 21). Such a regime develops "forms of normativity that define the conditions according to which imitations can be identified as exclusively belonging to an art and assessed, within this framework, as good or bad, adequate or inadequate: partitions between the representable and the unrepresentable" (21–22). The nineteenth century was the moment of consolidation of musical disciplines in the West, that is, of a particular type of institutionalization of music dependent on isolating normative principles for creating musical objects to be studied in particular ways. And this is one of the historical moments where the howlings of indigenous peoples make their phantasmatic and real appearance.

Guido Adler's 1885 foundational statement for musicology, *The Scope, Method and Aim of Musicology*, begins with the central role of the clear measurement of pitch in identifying organized tones as the essence of musical knowledge: "Musicology originated simultaneously with the art of organising tones. As long as natural song breaks forth from the throat freely and without reflection; as long as the tonal products well up, unclear and unorganised, so long also there can be no question of a tonal art" (Adler [1885] 1981, 5). One of the foundational texts for ethnomusicology, Alexander J. Ellis's *On the Musical Scales of Various Nations* (1885) is centrally concerned with the identification of pitch and scales of different peoples of the world as a foremost endeavor of the field. But the issue was not only the pitch-centeredness of the disciplines but also whether the favored simultaneity of the so-called common practice tonal system (harmony) was part of the given "natural" world or the product of a particular form of (Western) human creativity.

Questions regarding the relation between the physical properties of sound, the structure of consonance, the scales and human perception and musical creation, have been central to music theory in the West since Greek times to the present (Clark and Rehding 2001). How such questions are formulated and understood changes from one historical moment and place to another. What is particular about the late nineteenth century is how these questions were harnessed in the institutionalization and definition of the musical disciplines in Germany, still today defined as the foundational site for disciplinary formation of musical thought. This was accompanied by a new interest in the ear in the fields of medicine and acoustics, inspiring new practices of measuring sounds and discoveries about the ear's physiology and psychoacoustics (Sterne 2003). This gave rise to a renewed interest in "speculative harmonics" (Green and Butler 2002) that was central to the consolidation of the musical disciplines. The need for disciplinary foundations and a proper definition of music rearticulated age-old questions about musical origins and psychoacoustics, or what was then called the "psychology of tones" (*tonpsychologie*) or "the psychology of the folk" (*völkerpsychologie*). The latter fields have been considered the immediate antecedents of ethnomusicology and are often historically or thematically related to the anthropological questions about the notion of the people or *volk* associated with German Romanticism or, more specifically, with the German Counter-Enlightenment (Bunzl 1996). Not by accident, the scholars undertaking such experiments either worked directly with non-Western musics or referenced them in their speculations. After all, the issue of human and animal origins was central to the nineteenth century, and indigenous peoples had, for centuries, played a central role in the formulation of such a question.

For example, Eduard Hanslick's controversial theories of musical form and affect, which were influenced by Wilhelm von Humboldt and Jacob Grimm (Dahlhaus 1982), state that nature provides "material for the production of material, that is, of sound of high or low pitch; in other words the measurable tone. The latter is the primary and essential condition of all music, whose function it is to so combine these tones as to produce *melody* and *harmony*, its two main factors" (Hanslick [1885] 1891, 144; emphasis in the original). Both harmony and melody are not found in "Nature" but are rather "an achievement of man, only belonging to a much later period" (145). It is through "harmony that the first art emerged through utter darkness" (145). Rhythm is the only musical element found in nature, but it only becomes music by being tied to harmony and melody: "When South Sea islanders rattle with wooden staves and pieces of metal to the accompaniment of fearful howlings, they are performing *natural* music, that is, *no music at all* (emphasis in the original) (146). He affirmed that the apparent "naturalness" of the Western harmonic system was due to the "enormous spread of musical culture" and in a footnote added:

"the physically well-developed Patagonians of South America are entirely ignorant of both vocal and instrumental music" (147–48). For him, "the *physical* effect of music varies with the morbid excitability of the nervous system, so the *moral* influence of sound is in proportion to the crudeness of mind and character. The lower the degree of culture, the greater the potency of the agent in question. It is well known that the action of music is most powerful of all in the case of savages" (Hanslick [1885] 1891, 130). The penchant to enjoy the "physical effect" of music is felt by those, which are closer to Nature, "but music in this sense is not in the remotest degree enjoyed as a thing of beauty, since it acts as a brute force of Nature" (128).[8]

For Carl Stumpf, another founding figure of ethnomusicology, the line separating speech from song was initially enunciated through shouts that sounded like sustaining a long note, as in yodeling (Rehding 2000). The step needed for music to be created "was for two or more people to make a joint effort in order to increase the volume and, as if by serendipity, 'to discover countless polyphonic sonorities'" (Rehding 2000, 352). Stumpf related such ideas of musical development to his psychology of tone fusion.[9] Ultimately, the distinction between consonance and dissonance, for Stumpf, is an immanent element, based on musical categories of Western art music. The transformation from nonmusic to music is achieved through "cultivation," an idea that, Rehding says, is taken from Georg Simmel, a völkerpsychologie (psychology) scholar who was also influenced by Wilhelm von Humboldt's ideas. For Stumpf also, "what distinguished music from the 'sounds emitted by animals' . . . was its formal organization based on the interval, the variable difference in pitch between two sounds" (Ames 2003, 303).

Whether the harmonic system based on triads is a musical element found in musical nature, as affirmed by Stumpf and other German musicologists or constructed by the human mind, reaching its utmost development in Western musical development as implied by Hanslick, was a larger debate in the period that we will not address here.[10] But it does point out the centrality of pitch and harmony of the so-called common practice period in questions regarding the nature of music and of human nature. Such a polemic was also closely related to the question of musical origins.

The need to create a musical discipline defined primarily by the description of stylistic categories required a theory of origins based on music (Rehding 2000).[11] But, says Rehding, musicologists in the late nineteenth century did not search for an ur-music to represent such origins. Rather, they looked for a "first principle, the initial cause that made the historical progress of music possible in the first place . . . where origin was understood as that which is consistent in the face of change" (Rehding 2000, 346), in other words, whatever bears the trace of an identifiable, unchanging essence. The notion of origin invoked here carries an ontological connotation: "to know the origin of music, it was believed, means to know what music *is*," (emphasis in the original) conflating "historical research and metaphysical assumptions" (347–48). Ultimately, what contained the origins of music was an identifiable acoustic essence understood as an acoustic immanence because it remained traceable despite changes in musical history.

The identity of the comparative musicologist as a worthy scholar is thus based on his ability to identify pitch, while the identity of the indigenous peoples as (cultivated) persons with music is based on their ability to produce one. The methodology then requires finding a clear pitch and melodic line. In November 1885 Stumpf had the opportunity to hear a troupe of nine Bella Coola Indians at the Institute of Geography at the University of Halle. He was sensorially overpowered by a music that was impossible to transcribe due to the difficulty in identifying the pitch. So he sought a private encounter with one of the musicians called Nuskilusta. Eric Ames describes this encounter:

> They spent four evenings together in the Institute, working for one to two hours at a time. While Nuskilusta rehearsed the troupe's various songs more than ten times each, Stumpf plodded ahead with his hand-written scores ("LBI," 407). These private recitals differed in important ways from the public exhibitions; as Stumpf noted in his report, "Nuskilusta stopped rattling a piece of wood in his hand when he noticed that it disturbed me" ("LBI," 409). Such modifications could, he acknowledged, potentially distort the results of his experiment. "Nuskilusta kept the tempo slow, perhaps out of consideration for me, but also because [the solo performance] lacked the same effect that was produced by the collective singing and dancing" ("LBI," 408). During the day, he also attended the Bella Coola's public shows at a local beer garden, where he checked his revised transcriptions against the choral performances. Hours of intensive listening began to pay off, he observed, for "now I could hear more than mere howling; I could hear the melodies just as Nuskilusta had sung them *solo*" ("LBI," 408; emphasis in the original). The isolation and repetition of the individual singer's voice allowed Stumpf to train and adapt his aural response to the unfamiliar texture of Bella Coola music. If a melody could be picked from the "howling," then the strict separation between European art music

and natural music must be, Stumpf reasoned, at once wild and tenuous. (Ames 2003, 305–6)

Wild and tenuous indeed. Stumpf could produce, as a result of his labors, his seminal text *Lieder der Bellakula Indianer* (1886). But the impossible to transcribe howlings that remained beyond clear pitch perception posed another problem—they questioned not only the naturalness of Western art music, but also the validity of the stylistic categories used to define it. The repeated presence of howling vocalities from different parts of the world designated a potential "categorical crisis" (Fessel 2000) that needed to be solved if one was to admit that indigenous peoples have music and inscribe such vocalizations into musical description. Thus primitives as well as the sounds of nature seem to function as the limit within which to contain such a crisis but simultaneously as the elements that could easily destroy such "wild and tenuous" boundaries. In nineteenth-century musical studies in Germany such a distinction often took the form of what is now a classical anthropological denomination, the distinction between *naturalvölker* and *kulturvölker*.

Rehding says that the lack of archaeological evidence to prove musical origins was solved by a "methodological bias" of the period that made the naturalvölker into a people without history, and thus more tied to nature than to culture. Ultimately, says Rehding, "although a knowledge of the music of the Naturalvölker can apparently help us understand the origins of our own music, it is noticeable that the difference between "our" cultural music and "their" natural music is conceived, more often than not, as an unbridgeable binary opposition" (Rehding 2000, 358). But the story is more complicated because the problem is not so much that the naturalvölker do not have history but rather how they are integrated into one. This question is tied to the geopolitics and philosophy of studies of language in the eighteenth century in Germany by figures such as Johann Gottlieb Herder and Wilhelm von Humboldt. They were, in turn, highly influential on Heymann Steinthal and Moritz Lazarus, founders of völkerpsychologie, considered the direct antecedent field of comparative musicology and ethnomusicology.

According to Woodruff D. Smith, "völkerpsychologie was the comparative study of the characteristic mental patterns of different peoples, with particular emphasis on the historical development of those patterns" (1991, 115). One of Wilhelm von Humboldt's major contributions to the theoretical underpinning of this enterprise is his formulation of the relationship between history, comparative anthropology, and folk psychology. Following the Herderian legacy of the "German Counter-Enlightenment" (Bunzl 1996), for von Humboldt history was meant to follow the cosmopolitan ideal of the recognition of the Volk as the key to the diversity of each society. Comparative study of the specific traits of each Volk would yield the general history of humanity. For von Humboldt, "each individual *Volk* had a *Nationalcharakter*, a distinct *Volk* character, which was embodied in the totality of its outward manifestations: traditions, customs, religion, language and art. These in turn revealed the degree of *Bildung* attained by a given nation. . . . However, some nations, including the Germans, English, French, Italians and the Greeks had made the most of their innate potentialities and reached a higher state of self realization" (Bunzl 1996, 22), though the expectation was that each national character would eventually reach such a state. His "plans for a comparative anthropology" thus involved empirical observation of different traits with attention to "historic detail," detail that was furnished by tying history to the study of empirical data. Such data were gathered by paying attention to "physiological factors, which operated on the exterior forms of all living entities, such as individuals, nations, peoples, and the entire human race, with discoverable regularity" (25). But also, and more importantly, by accounting for "irrational psychological factors such as 'abilities, feelings, dispositions, and desires,' which were inherent in the agents of history and [unlike nature] completely eluded 'discernable laws'" (26). The goal of the historian was to understand through induction those texts that yielded the psychological character of a people and the context from which that character emerged. Völkerpsychologie was to tie this into larger units by discovering the general laws or "psychological essence shared by all members of a *Volk* and the driving force of its historical trajectory" (28). The objective was to study concrete manifestations of the "psychological products of a people, foremost its language and mythology, but also its religion and customs" (28) and, of course, music, in order to find the laws that governed psychological development and map them over space and time.

Thus the humanist distinction between a natural culture and a civilized culture, as presented by Rehding, is complicated by the idea of the character and psychology of a people being temporally and spatially codified through the study of its artifacts—language, customs, traditions, and sonic utterances. The move is one that seeks to psychologize language, customs, and traditions in search of a character. While the products of civilized culture are mapped onto aesthetic ideals of civilized beauty, the products of Volk character are differentially mapped onto sensorial and trait generalizations about national psychologies. These two contrasting notions of cosmopolitan ideals—one that seeks to generalize through a universal distinction between culture and nature, and one that seeks to generalize by recognizing the diversity of national characters while searching for comparable general laws—map onto each other by generating an ellipsis between culture and character through the empirical artifacts used to produce such characterization—language, vocalizations, customs. Primitive music, with its heightened sensorial dimensions, can easily be made amenable to Western constructions of nature rather than art, to psychological essences rather than singularities. If one looks for acoustic essences, the other looks for formal traits that establish the essence of peoples. The relation between both yields the Western folk notion of a particular musical tradition as characterized by particular traits that in turn characterize a people. If ethnomusicologists were to look for conventions characterizing nations, musicologists were to identify the traits of a particular singularity, framed either as authored works or as a collective achievement yielded by the summation of such inventions.

Much of this discussion on historicism, language, nature, and human nature is the product of a German eighteenth-century critique of "universal Enlightenment" and is referred to as the "German Counter-Enlightenment" by some (Bunzl 1996) or as an "Enlightened Vitalism" by others (Reill 2005). The Humboldt brothers were crucial to the articulation of this critique (Reill 2005). However we place these ideas within a European discussion of the relations between multiple eighteenth-century notions of Enlightenment, what is interesting for us here is the centrality of colonial Spanish America. This German discussion was not only crucial for the formation of the ideas about culture and language, via such figures as Herder, Wilhelm von Humboldt, and Franz Boas (Bunzl 1996). It was also important for the formation of ideas about nature and human nature through such figures as Georges Louis Leclerc, Comte de Buffon (1707–1788), and Alexander von Humboldt (Reill 2005). Comte de Buffon was avidly read by naturalists in New Granada while Humboldt based his work on an extensive trip through the region during which he learned from scholars in the Americas, drew from the interpretive traditions of the region, and used the data of colonial bureaucratic archive to formulate his theories (Cañizares Esguerra 2006). While Wilhelm developed his theories in relation to scholars working in Spanish America, Alexander was once hailed as America's second "discoverer" and the polemic of his Latin Americanism versus his Eurocentrism has been a central topic of debate among historians of science (Nieto Olarte 2010).

As is evident by the discussion above, central to the definition of music were the types of sounds considered musical and human and the practices used to establish distinction between natural sounds and those produced by humans. If the Europeans were busy elaborating a theory of art, is it possible to at least speculate, on such scant historical evidence, how the bogas were thinking about the "doing and making" (what we call art) of their vocalizations? What if we consider for a moment that the bogas could have purposefully sounded like animals because they were deliberately imitating such sounds in their vocalizations? In the rest of this chapter I will elaborate on the questions of ecology of acoustics raised by the problematic boundary line between musical sound and sounds from nature raised by the fact that the bogas sound like animals and incorporate elements from the nonhuman world into their vocalizations in multiple ways. I will do so by contrasting an alternative interpretation of the sound of the bogas with the understanding of sounds of nature in Alexander von Humboldt's work. I begin with the bogas.

On the Ecology of Acoustics

As stated by Claudio Lomnitz, "The conquest . . . was a process by which people and things were arranged into new classes, people were addressed as new subjects, and old subjects ceased to be addressed as people" (2005, 65). The question that emerges is how do the peoples who are the object of such drastic destitution reconstitute a sense of continuity, collectivity, belonging, and creativity in the midst of massive destruction and forceful transformation? Marilyn Strathern defines collective "as a form of activity in which persons come together on the basis of shared characteristics. What they hold in common is regarded as the rationale of their concerted action" (1988, 48). But in situations of great upheaval as happened in colonial Spanish America, how do forms of concerted action emerge and how are the collective and the singular reconstituted? Since such a question implies taking into account both human and nonhuman worlds and sounds, then the question inevitably leads us to ask how, in such a situation, one reconstitutes an acoustemology or "the world sonified" as a world that is "known," "felt," and "performed" (Feld 2012, 131). Such issues concern not only the colonial history of music and other expressive genres (Baker 2008; Hanks 2010) but also the politics of "the acoustics of ecology and the ecology of acoustics" (Feld 2012, 125).

Paul Carter has proposed a political ecology of sound that highlights the ambiguity present in auditory knowledge "as one that is constitutionally environmental and situational. It corresponds to the participatory, or echoic, production of meaning" (Carter 2001). Building on Roy Wagner's notion of echolocation, developed from the way bats locate themselves in the world, he proposes sound as providing a form of orientation in the relation between entities in the world that builds on the impossibility of resolving the communicative ambiguity implicit in the relation between sound and sense in the notion of the sign. Carter applies the notion of "'echoic mimicry' to situations of cross-cultural colonial encounter" and to migrants, "in which the concept of origin ceases to have value, being replaced by a notion of beginnings repeatedly begun" due to "the collapse of an entire auditory topography." This entails, according to Carter, the creation of an "in-between" sound situation that implies the reconstitution of sounds in a process of environmental echolocation of the parties involved in the exchange of sounds and listening through a process of creative mishearing, "an acoustically-shaped place-making" (Carter 2001). According to Carter, this requires more an orientation toward the other, a disposition to listen, rather than a resolution of the ambivalence involved in such a situation leading us away from a politics of representation and essentialism to one of echolocation that highlights ambiguity as central to the reconstitution of the world in colonial situations. He thus contrasts the politics of listening of colonial encounters which semiotically close off ambiguity with a politics of echolocation based on the ambiguity of mishearing as central to the mutual reconstitution of place and one's own place in the world through sound even in drastic situations.

While I agree with his move away from sound as communicative transparency that emphasizes difference as a politics of representation, and with echolocation as a means of rearticulating a new auditory and expressive world, I hesitate to apply a diplomatic possibility even to notions of echolocation in situations of drastic power imbalance like the colonial one. If anything the history of Latin America and the Caribbean teaches us how politically complicated such in-between and ambiguous transactions are. But Carter's emphasis on echolocation and his radical questioning of the politics of representation as the site of colonial disjuncture as well as the distinction between human and nonhuman sounds to explore the politics of an "acoustics of ecology" is crucial for rethinking the acoustics of the colonial. I will use his ideas in a speculative exploration of the bogas as they "were listening to the production and reproduction of categories, listening to the ordering of things" (Feld 2012, 131), as well as in exploring Humboldt's own modes of listening as crucial to developing his understanding of nature. I intend to explore the ways the bogas might have been participating in a "history of human listening and eco-acoustic evolution" (Feld 2012, 125) central to the reconstitution of the politics of life in the reconstitution of a sense of the collective. This is an adventurous proposal due to the scant historical acoustic material presented in the earlier part of this chapter on which such speculation is based. But it is worth attempting, more because of the possible questions it raises about the topography of sounds in colonial situations than about the questions it answers about the bogas. I begin by exploring the history of emergence of the bogas from the sixteenth to the eighteenth centuries, in order to provide a historical background for such speculation.

COLONIZATION AND THE RECASTING OF THE COLLECTIVE

The boga of the Magdalena River, as the institutionalization of boat rowers as a particular labor class[12] and type of service eventually came to be called, was initially created through the illegal, forced work of indigenous people (Noguera Mendoza 1980). During the sixteenth century, Mompox was the most important riverine port on the Caribbean portion of the Magdalena. The colonization of the indigenous groups of the river in the area around Mompox was undertaken through an *encomienda,* an institution of servitude of the Spanish colonial empire in which the King gave his Spanish subjects the right to subjugate the Indians in order to receive tribute in return for the supposed responsibility of protecting and missionizing them. In this case, the encomienda forced the Indians into the slave labor of nonstop rowing in the tropical heat of the Magdalena's famously turbulent waters.[13]

By the eighteenth century, those working the boga were primarily free *zambos,* men of mixed indigenous and African descent, employed by a patron who owned the boat. The boga could not be manned by the labor provided by enslaved Africans because the seven to eighteen rowers of a typical boat in the Magdalena River, which carried a similar number of white passengers, could easily abandon their stations, leaving the passengers to the mercy of the unknown and vastly unpoliced geography of the Colombian Caribbean. By then, the bogas had a virtual monopoly over transportation in the Magdalena River (Peñas Galindo 1988).

Colombia has the third-largest population of African origin in the Western hemisphere after Brazil and the United States, and "the Caribbean Coast is the region of Colombia that is most densely populated by people of *mixed* African descent" (Helg 2004, 2; emphasis mine).[14] The 1777–80 census pointed out that "free people of color [libres de todos los colores] represented between 89 and 100 per cent of the inhabitants in most villages and small towns located along the Magdalena River (including Barranquilla, then a parish of 2,934), in the Plain Northeast of Cartagena, along the San Jorge and Cauca rivers, and near Riohacha" (Helg 2004, 43). For Aline Helg, "the *libres de color* in the entire region between Cartagena, Barranquilla and Mompox as well as on the riversides of the Magdalena and Cauca rivers were almost exclusively of *mixed* and full African ancestry" (emphasis mine) and were scattered throughout this vast territory (Helg 2004, 43). If they were mixed, they obviously had other ancestries as well.[15]

Even though the interpretation of the data of this census can be disputed by questioning the practices of documentation, the different modes of classification of peoples in each of the provinces, and the practices of racialization of the period,[16] the point I wish to make remains valid: in the Colombian Caribbean it is difficult to reduce the constitution of socio-racial differences and hierarchies during the colonial period to a simple opposition between blacks, indigenous peoples, and whites. In New Granada in general, 51 percent of the population was deemed of mixed origin in the eighteenth century.

Aline Helg speaks of the "racial fuzziness" of Caribbean New Granada (2004) and authors such as Mosquera, Pardo, and Hoffman (2002), among others, speak of the "complex" sociopolitical relations that make it difficult to reduce the networks of association and resistance, survival, creativity, and adaptation of Afrodescendants to simple oppositional categories. In societies based on distinguishing between Creole elites and commoners through identification of origins, such fuzziness yielded a governmentality obsessed with legal practices to establish such distinctions through "identifying and enforcing racial hierarchies" (Cañizares-Esguerra 2006b, 38). In eighteenth-century Colombia, access to political rights and educational opportunities depended on demonstrating blood purity (Castro-Gómez 2004). The social and political hierarchies generated by American genocide and the Atlantic slave trade were crucial in the construction of capitalism, Western European expansionism, and racialized global constructions of knowledge and power (Quijano 2000). But this does not mean that the notions of personhood, the practices of knowledge or the politics of relationality between peoples were delimited solely by the politics of racialization:

> In all of the analyses [or essays in the book they edit and introduce] the profiles of Afrodescendants become blurry when faced with the complexity of concrete situations: unexpected alliances, tactical coincidences and fateful encounters demonstrate the capacity of Afrodescendants of acting, with or against their neighbors, to maintain and adapt themselves to the historical and geographical contexts to which they were taken. There comes a moment when the categories commonly used to qualify people and social groups ("whites," "blacks," "slaves") do not attest to their complexity and become, in and of themselves, an object of discussion and negotiation, as much for the observers as for "Afrodescendants" who frequently do not name themselves in such

ways . . . the roads of identity of Afrodescendants are in reality much more complex and tortuous than a simple opposition between black and white. (Mosquera, Pardo, and Hoffman 2002, 24)

On the other hand, Arias and Restrepo (2010) state the need to historically contextualize words that today we tend to associate with race such as "white," "black," or "blood purity" and to distinguish between the historical uses of those terms in different circumstances and the conceptual and analytical work we demand of them in the present. Also, such terminologies were not used in the same ways by the elites and subaltern sectors of the population: "it is pertinent to consider that the systems of racial classification that operate in the subaltern sectors are not simply the projection of the mechanical appropriation of those elaborated by the elites nor vice versa" (Arias and Restrepo 2010, 60–61). Thus, to pretend that peoples who came from Africa and indigenous peoples of the Americas or the mixed peoples who inhabit the continent exist solely as "racial" constructions of the West is, paradoxically, to undertake the ultimate colonial erasure (Viveiros de Castro 2011). So, while the terminology I am using here is the one historically used to classify peoples in the documentation of the period, this does not mean that the notions of personhood and modes of relation were restricted to the elites' means and boundaries of differentiation. I want to turn now to the broad patterns of relation between elites and commoners in the region, especially, in our case, those of *free peoples of all colors.*

In rural areas, the small, white Caribbean elite lived by cattle ranching, legal or illegal trade, or work in regional agricultural markets. But, unlike other places in the Caribbean, the region never developed a large plantation economy during the colonial period, despite the strong presence of *hacendados* (landowners), until the banana plantations of American imperialism in the early twentieth century (Helg 2004). There was a very scant presence of the Catholic Church and the colonial state in small villages and in rural regions, which meant that family structures, sexual unions, and labor practices were either unpoliced or policed through the patronage and clientelage practices of the hacendados. In places where the colonial state and church had a stronger position (as during the Inquisition in Cartagena in the seventeenth century), the types of associations of African nations that existed in Cuba, for example, were repressed (Maya Restrepo 2005). As such, forms of resistance included escape to the hinterland, the formation of *palenques* (rebel enclaves of escaped slaves), and associations of "black sorcerers," which maintained alternative religious and cultural practices throughout the region (Maya Restrepo 2005).

These sites acted as dispersed, effective spaces of resistance. And when these groups were militarily defeated, the survivors of palenques and of indigenous groups had the option of moving to another unpopulated region, due to the unpoliced nature of the hinterland and the ineffectively small police forces (Landers 2002).

Part of the Caribbean has historically consisted of a large frontier region — marked by the Guajira peninsula toward the east and the Darién toward the west, and to the Orinoco-Amazon basin to the southeast — that was characterized by prolonged indigenous resistance, transnational contraband, and a constantly shifting population of runaway slaves and of free peoples of all colors (Serje de la Ossa 2005). While the frontier lands of the Sierra Nevada de Santa Mata, Darién, Orinoquía and Guajira were settled by indigenous groups in either contentious or negotiated relations with other peoples, the population around the Magdalena River and in other areas of Colombia, such as the Cauca River, was primarily made up of free peoples of all colors (Colmenares 1976). While some of these populations sought their formal recognition as towns, others were constituted as *rancherías*, spaces outside any formal jurisdiction, which were, nevertheless, prone to repeated attacks in order to bring them under control (Garrido 2007) or to the changing forms of association and contention between different groups of commoners. During the second half of the eighteenth century there was an expansionist campaign of landowners and cattle ranchers in the area around Cartagena with forced practices of mobilization of populations. Yet simultaneously, the free peoples of all colors tried to establish, through legal procedures and different tactics of land population and mixture, control of small, independently owned areas of production (Sánchez Mejía 2011).

At the same time, the population of white urban elites was quite small and divided due to different forms of factionalism and political strife.[17] One of the consequences of this factionalism was that no strong regional political project ever coalesced in the Caribbean (Múnera 1998) although this factionalism of the elites has also been ascribed to a Humboldtian socio-racialized differential politics of geography (Serje de la Ossa 2005).

After the seventeenth century there was no *massive* importation of slaves to the Colombian Caribbean, and by 1800 there were hardly any African-born peoples in Colombia (Helg 2004). This made it more difficult to construct south-south transatlantic cosmopolitanisms and processes of identification between Africa and Colombia in a late colonial or early republican period as happened, for instance, in Brazil (Matory 2005) or in Cuba (Palmié 2002). Therefore during the colonial period, African ancestry did not become "an identity for political organization" (Helg 2004, 14) in this area. But even in clandestine *cabildos* or palenques during the colonial period, associations were not necessarily restricted to blacks.[18] "Popular mobilization" was then difficult in such a "multiracial society where slavery was only one of several labor systems" (Helg 2004, 14). Despite the existence of repeated, organized rebellions of indigenous peoples as well as slaves, from the sixteenth to the eighteenth centuries, the strategies of resistance were generally fractured and dispersed and did not coalesce into massive revolt.

Some of these strategies included overt individual or group flight from slavery; outright insurrection (even if it was small scale and/or in association with indigenous groups); the formation of rebel towns or palenques; incursions into cities to attack a particular slave owner or site, or to free other slaves; the flaunting of family mores through different types of sexual liaisons, mestizaje in the sense given to it by Elisabeth Cunin, that is, as a form of "conscious usage or management (*gestión*) of alterity in which the other is given a changing and polysemic status" (Cunin 2002, 281); legal struggles for recognition through debates about honor that challenged the caste system (Garrido 2007); indigenous takeover or colonization of territories that began to be considered as hinterlands of the nation (Serje de la Ossa 2005); and, very importantly, in the case of slaves, the buying back of freedom. The latter practice became so common that by 1851, when slavery was formally abolished in Colombia, most slaves were already part of a large population of "free people of all colors" (Mosquera, Pardo, and Hoffman 2002). The free peoples of all colors from the Caribbean, in contrast to those from other regions of Colombia, were particularly rebellious. They resisted living "*en policía*" (under official command of church or state) and often lived in the large interior frontierland where we also find an indigenous population that resisted their reduction to the governmentality of Christian missionaries (Garrido 2007).

Communication and transportation between the Caribbean and Andean regions was often very difficult and possible only via the champanes in the river, since travel by land was close to impossible in many areas and often came to a complete standstill during the rainy season. As stated by Helg, in 1789 "most of Caribbean New Granada resembled a patchwork of scattered Indian pueblos and black, mulatto and zambo villages surrounded by hacendados' fiefdoms on a background of unconquered lands" (2004, 42), an image that is reinforced by reading the travelers' accounts. The bogas were well aware of the strategic advantages of these characteristics: "once on their way, to the great displeasure of their passengers, they stopped as often as they could to fish or gather eggs, or to drink or have sex with the riverside population" (Helg 2004,

68). As expressed by Humboldt in his travel diary, "one is a slave of one's boat rowers" (Humboldt 1801, 32).

The zambo bogas' status as free men was inseparable from their position as the only labor solution to the problem of transportation of peoples and commodities for the colonial empire and the early postcolonial nation. If the bogas seem to be resisting the system in their independence and flaunting of authority during navigation, ultimately they were also a key component of its constitution, one that they had no opportunity of changing. In fact their ostentatious behavior came to be used by Andean elites as proof of their lowly animal condition: "the boga, descendant of Africa, and son of races debased by tyranny, has of the human only the external form and the necessities and primitive forces" (Samper 1862, 27). They thus occupy a highly ambivalent position: free men, a needed labor class, yet the most racially debased persons in New Granada. In no small measure their debasement was based on their ostentatious behavior, which made it impossible for elites to subsume them under the compassionate benevolence of enlightened liberalism in a politics of folklorization of their expressive gestures, and on their ambivalent position as "free men of all colors."

As stated by Elisabeth Cunin, what worried the authorities was not so much the slave but the mestizo, the free of all colors (2002a). For Cunin "the free of color contests the social order—neither master nor slave—and the racial one—neither white nor black—due to the intermediate position they occupy" (2002, 14). Mestizaje in colonial America, rather than overcoming racial differences as proposed by different ideologies of racial democracy in early twentieth-century Latin America, was perceived as a threat (Cunin 2002). But this does not mean that by that token the categories of black, white, or Indian were reinforced. In making it difficult to objectify hierarchies of difference, mestizaje "impedes the edification of a clear frontier between us and them" and "questions any classification of a clearly defined identity" (Cunin 2002a, 289). In doing so, it questions two modes of understanding: what Cunin calls "the analytic . . . the decomposition into pure elements (the white against the black of multiculturalism) and the synthetic . . . the reconciliation of contraries the myth of racial harmony, or assimilation)" (290). Moreover, the categories of mestizos and free of all colors carried a heightened ambiguity because they were founded on the biological, the sociological, and the cultural. Such conceptual "hybridity" (De la Cadena 2000) and ambivalence (Wade 1993; Cunin 2002, 2003) was not only used by the elites in heightening racialization. It also allowed mixed subalterns to develop multiple political, social, and creative strategies, as seen in the previous section and as is evident in the way the bogas handled their boats. Because of this, Latin Americanist anthropologists in Latin America have challenged the idea of mestizaje as mixture and proposed alternate meanings.

Cunin, writing about Afrodescendants in present-day Cartagena, proposes a notion of mestizaje as a practice that "does not mean 'mixture' of 'hermetic' cultures but rather the negation of the logic of isolation and separation itself, of original purity of cultures in contact" (2003, 17). Here, mestizaje is not thought of "as the result of the encounter between European, indigenous and African populations, but rather as a point of departure that turns all search for origins into an illusory endeavour and all notions of culture discontinuous" (Cunin 2003, 27). For her, "mestizaje is not a state or a quality but is rather situated in the order of action" (Cunin 2003, 27). Thus, she speaks of a mestizo competence that emerges in the interaction between individuals: "I am interested in the way members of a society classify themselves and are classified according to their physical characteristics as they interact. This leads us to an analysis of the capacity of individuals to know, mobilize, and apply the rules and values that are proper to each situation, of moving from one normative frame to another, to define their role and that of others in an interdependent way; that is what I call mestizo competence" (Cunin 2003, 8–9).

Elisabeth Cunin's notion of mestizaje as a competence emphasizes the "dynamic and interactive mechanisms of elaboration of the frontiers between 'us' and 'them,' or, in more general terms, of emergence and fixation of social norms" (20). This logic of mestizaje as a conscious interplay between being and becoming is a way of understanding the conscious and historical usage of multiple knowledges in the mobilization or practices of self-definition. The status of the bogas as both a free subaltern mixed race group in a slave society and a member of a guild with the specialized knowledge for handling and, in a way, controlling the movements of goods and people made of them a group of people who through their labor, that is, *as a working class*, were simultaneously inserted into several overlapping networks: that of their patrons, that of the riverine population, that of the elite, that of contraband and other "illegal" activities, that of the relations with other *libres de color,* that of the different populations they encountered throughout their travels. As such, they were often at the crossroads of various "contradictory orders and levels of obligation" (Povinelli 2002, 2) of the colonial empire and the emerging nation but also clearly positioned to learn and articulate different understandings of convention and creativity, of the given and the made, from different peoples.

In the history of conquest and strategies of survival and resistance discussed above, we see that, from the sixteenth to the eighteenth centuries, entities that previously were not part of the collective texture had to be taken into account in the reconfiguration of social life. The production of new forms of design of the acoustic and practices of its inscription was central to this transformation. The question is how does one conceive of the new forms of reconstitution of the vocal in the politics of echolocation of the bogas as a history and historiography that permits the "reconstruction of a series of events (at the psychic, socio-historical and technical level)" (Lévi-Strauss 1966, 33) through the reordering of acoustic "design" (Ingold 2011). This is not to say that the "cultural" resolves the contradictions of the political but rather an attempt to understand the politics in the emergence of new forms of vocalization. If we understand descent and inheritance as openness to endless variation, as a possibility of enhancing becoming, then it is possible to unhinge a politics of identity from a melancholic search for repetition inherent in the politics of representation (Grosz 2011). In this sense, then, "the past is not the causal element of which the present and future are given effects but the ground from which divergence and difference erupt" (Grosz 2004, 8).

Clearly, the bogas were positioned to move between different orders of obligation, peoples, and entities. In her study on what the Inquisition[19] called "wizardry" (hechicería) and "witchcraft" (brujería) and which she calls "magical practices," Diana Ceballos explores different forms of rationalization, systems of knowledge, and symbolic order in the late colonial period in New Granada (1994, 2001). These practices, denounced as crimes of idolatry during the Inquisition, were ascribed to indigenous peoples, Africans, free peoples of all colors, conversos (Jews and Muslims converted to Catholicism), and poor whites, but were not solely used by them. They were also used by "cultural intermediaries," people located at the nexus of several orders of obligation—health practitioners, artisans, militiamen, servant women, those who administered love potions, bogas—and by members of the elite who required such services.

What is interesting about the seventeenth-century Inquisitorial archives is the detail by which they differentiated multiple practices. Ceballos shows the different practices mentioned: witchcraft, demonic witchcraft, wizardry, shamanism, yerbatería (use of plants with a particular end), herbolaria (profound and systematic knowledge of nature), curanderismo (use of witchcraft to cure or to make sick), ensalmos (religious chanting). I will not define each of these but simply call attention to the fact that they all involved handling knowledge about and between human and nonhuman realms, be they from the "natural" or "spiritual" worlds. All of them also involved active transformations of forces of "alterity" (however they were specifically conceived), through use of herbs, vocalizations of different sorts, embodiments of those alterities, and appropriations of some of their qualities in order to transform or affect some aspect of themselves or of others. Thus, while clearly establishing a differentiation between "us" and "others," such a distinction did not involve a practice of epistemological purification in order to separate (Latour 1993) but of the active transformation of alterity through different modes of crossover (to use a musical term) between realms of beings.

Ceballos explains that in the sixteenth century some of these practices were recognized and described in detail but not necessarily negatively; in the seventeenth century they were criminalized by the Inquisition, but by the eighteenth century, the accusations practically disappear.[20] Ceballos sees such transformation as one of "transculturation" in which practices formerly assigned to peoples of particular origins were absorbed, by the eighteenth century, into a more common "popular culture," generating what she calls "typically [Latin] American phenomena" (Ceballos 1994; 2001).

Adriana Maya, researching African practices and identities in New Granada in the seventeenth century, also identifies the *juntas*, or nocturnal meetings done by criollos, mulatos, and zambos in urban and rural areas, as particular sites of resistance, which she labels as *cimarronaje simbólico* (symbolic maroon practices). She sees these as sites of "resignification" of what she calls "African corp-orality" (2005), using techniques of bodily and vocal expression, objects from nature, and invocation of dead spirits. Besides the practices mentioned above, she mentions practices of divination, different ceremonies, and initiations. Her detailed, historical material offers more evidence of the modes of mobilization of practices between human and nonhuman relations. Ultimately Ceballos and Maya Restrepo, the one emphasizing transculturation and the other African retentions, take a drastically opposite approach to interpretation of this history. The debate between the "strategic" uses of contemporary Afrodescendants in claiming their "identities" and the Herskovitzian stance that sees in certain practices traces of Africanity (*Huellas de Africanía*) has been particularly intense in Colombia in the past two decades. Currently such a debate is being channeled through a historically positioned critique of the emergence and uses of the notion of race in different contexts and the need to rework our understanding of colonial history through rethinking the history of the relation between nature and culture.[21] In what follows, I propose a different historical reading by relating the contemporary ethnographic literature with the historical archive. I seek to do so in order to explore modes of opening questions about the acoustemologies of the colonial to alternate readings.

The eighteenth century, then, seems to be a particular moment of consolidation of relations between different subaltern populations through mutual exchange of practices that became part of the network of relations between peoples. This coincides with the fact that the drastic population decline that had been unleashed by the conquest in the sixteenth and seventeenth centuries abated, and both indigenous and mixed populations began to grow again in the eighteenth century (Melo 2011), something that also led to a reorganization of indigenous groups (Langebaek 2009). We can then presuppose that, far from being erased, these practices became part of the texture of the everyday generating an ambiguity that allowed for multiple interpretations and uses of legacies within such a historical context. So we can also read this history in relation to the contemporary ethnographies of the region linking elements that are present in both indigenous and Afro-Colombian practices.

Ethnographies of music, myth, and/or narration in northern South America's aboriginal groups point to the practice of either imitating or making reference to the sounds of birds and other animals as a component of musical expertise and a common feature of the sonic dimensions of the everyday.[22] Although the resulting musical forms of speech/music genres do not necessarily imply imitating such sounds (Seeger 1987), the capacity for imitating them or for learning songs from animals, is repeatedly stated in these ethnographies. Also, in many Amerindian narratives of origins of music or of processes of learning music, music and speech genres tend to come from what Seeger (1987) calls "the outside" (i.e., the nonhuman realm, be it mythical beings, foreigners or animals). In Afrodescendant ritual practices as well, both drums and peoples have the capacity of envoicing nonhuman (spiritual) entities. This does not mean that the distinctions between self and other disappear. But such a distinction does not imply the use of the same ground between nature and culture as that which created the naturalvölker and the kulturvölker (Viveiros de Castro 1996). Rather, what we see here is an understanding of alterity as a means to a transformative process and not as recourse to instantiate separability as a process of purification.

If sounding like animals, learning sounds from animals, or incorporating nonhuman entities in sound is not a problem but an objective, then it becomes evident that the human-nonhuman relation, or the relation between nature and culture present in the voice is not one that debases the person, as explored in the previous section. The type of "relational ontologies and their acoustemologies" raised by such understandings of the voice are rather ones that are used to link "place to cosmology through sound" (Feld 2012, 126). This brings to the center questions regarding the relation between acoustemology and place that has been crucial to the anthropology of music (Feld 1996, 2012; Seeger 1987; Menezes-Bastos 2007). It also requires exploring an ontological difference in the understanding of the human and nonhuman, a classical theme in anthropology that has gained intensified currency due to the irruption of "nature" on the academic scene and the consequent need to rethink the notion of the person and of different forms of relationality and kinship. This has implied an intense revisionism of the idea of animism inherited from the nineteenth century.

As noted by Sahlins, the ontological difference between nature and culture that has prevailed in the West since roughly the seventeenth century tends to be quite exceptional: "As enchanted as our universe might still be, it is also still ordered by a distinction of culture and nature that is evident to virtually no one else but ourselves" (Sahlins 2008, 88). In principle, for many cultures of the world, the common condition of animals and humans is not animality but humanity (Viveiros de Castro 2010; Sahlins 2008) with the consequence that there are "inverse semiotic functions attributed to the body and to the soul" (Viveiros de Castro 2010, 28). One way of approaching the implications of this for understanding the vocalization of the bogas is to use the reconceptualization of animism that this implies. The concept of animism was originally formulated by Tylor in 1871.[23] Throughout the twentieth century it was generally thought that animism entailed the idea that certain cultures attributed a soul or divinity to specific beings of nature, in an anthropocentric extension of the Christian notion of the soul. But it is precisely this anthropocentric projection that has been questioned. Tania Stolze Lima, in her ethnography of the Juruna from the Amazon, summarizes the problem:

> A proposition such as "the Juruna think that animals are humans," besides deviating appreciably from their discursive style, is a false one, ethnographically speaking. They say, "The animals, to themselves, are human." I could, then, rephrase this as "the Juruna think that the animals think that they are humans." Clearly, the verb "to think" undergoes an enormous semantic slippage as it passes from one segment of the phrase to another. . . . the Juruna could tell us: what you consider to be human characteristics (as you define them both naturally as well as metaphysically), do not belong intrinsically to human beings. We have to produce these characteristics in ourselves, in the body. As we shall see, anyone—animal or human—may produce the characteristics that best please themselves. (1999, 113)

In principle then, what is common to animals and humans is the capacity to think of themselves as social collectivities, as having homes, undertaking rituals, singing, and so on. Thus the capacity of the bogas to move between the world of the human and the nonhuman by envoicing animal sounds is not a "lowly condition of animality" but rather is due to the shared capacity of humans and animals to have a voice, sing, and speak. Songs and languages can potentially be transmitted between species by entities who have the ability to do so. If this is so then having a particular type of voice does not imply representing a single entity since potentially different species could use that same voice or transmit it between them, even if they do not conceive of it in the same way.

However, the perspective according to which each species conceives of this voice is different. If animals conceive of themselves as singing and having voices, it does not mean that all beings share the same point of view: "numerous peoples of the New World (very likely, all) share a concept according to which the world is composed of a multiplicity of points of view: all existents are centers of intentionality, that apprehend other existents according to their respective characteristics and capacities" (Viveiros de Castro 2010, 33). Thus, "a similitude of the souls does not imply that these souls share what they express or perceive. The way that humans see animals, spirits and other cosmic actants is profoundly different from the way that those beings see them and see themselves" (35). This "perspectivism" or "multinaturalism" (Viveiros de Castro 1992, 2010; Stolze Lima 2005) resides in the differences in the bodies thought of not so much as "physiological functions" but rather as "effects that singularize each species of body, its forces and weaknesses: what it eats, its forms of moving, of communicating, where it lives, if it is gregarious or solitary, timid or arrogant" (Viveiros de Castro 2010, 55). The body as a "bundle of affects and capacities" is "what lies at the origin of perspectives" and is what permits the generation of "relational multiplicities" (55). So the fundamental fact, let us say, is not that the bird thinks of its bird song as a song in a ritual feast and the person thinks of that same bird song as simply the sound of a bird. This would be a cultural relativism in which the idea of culture is simply extended to other species. Rather, the sonorous object ritual song/bird sound does not have an essence but is conceived as a multiplicity through which a relation is constituted—as such alterity is *inherent* to things or, in this case, to specific acoustemes. Alterity is thus understood "as a condition of the possibility of being" (Sahlins 2008, 47). Multinaturalism is not so much "a variety of natures" (applying the notion of relativism to nature) but rather "variation as nature" (Viveiros de Castro 2010, 58).

All theories of art, or of design with particular forms (Ingold 2011), imply a relation between a being that perceives and a perceived object. So we also need to think not only about the understanding of the vocalizations in question but also of the person who is listening in order to understand the relation between them. Marilyn Strathern had already questioned the idea of using our own division between "the individual" and "society" as the means to undertake an analysis of the relation between the person and the collective in different cultures. Rather, she places an emphasis on a notion of personhood, understood as a movement between states of multiplicity or unity, manifested either singly (in a single person) or collectively (Strathern 1988). Thus notions of relationship across difference depend on how the relation between personhood and alterity itself is conceived. Such notions of alterity are often understood through the relation between structures of kinship and the cosmos where "alterity is a condition of the possibility of being" (Sahlins 2008, 48). Sahlins says, summarizing the ethnographic literature:

> Ethnographic reports speak of the "transpersonal self" (Native Americans), of the self as "a locus of shared social relations or shared biographies" (Caroline Islands), of persons as "the plural and composite site of the relationships that produced them" (New Guinea Highlands). Referring broadly to the African concept of "the individual," Roger Bastide writes, "he does not exist except to the extent that he is 'outside' and 'different' from himself." Clearly, the self in these societies is not synonymous with the bounded, unitary and autonomous individual as we know him—*him* in particular, as in our social theory if not our kinship practice. Rather, the individual person is the locus of multiple other selves with whom he or she is joined in mutual relations of being; even as, for the same reason, any person's self is more or less widely distributed among others. (Sahlins 2008, 48)

Let us return then to the bogas and explore some ideas about vocality based on such notions of the relation between a self that is not conceived as an autonomous being but as the locus of a "transpersonal self" and a vocalization that is conceived as a multiplicity rather than embodying a sound that represents an entity. In the first place we could say that the voice is not so much a mechanism that permits the mediation between the world of humans and nonhumans and between the signifier and the signified (Dolar 2006), but instead it permits the manifestation through en-*voicing* (in-vocation) of relational multiplicities—a capacity to manifest "bundles of affect" of the type, for example, song of a ritual feast/bird sound, that imply different things for the different entities that produce or hear them. Voice thus permits the "sharing of certain attributes" (Sahlins 2013, 31) between beings where relations between entities are conceived as constituting a "mutuality of being" (Sahlins 2013). Voice rather than a mediation between worlds is "a medium of mutuality" (Sahlins 2013, 54) in the constitution of a notion of a distributed self. Second, the idea of an acoustic mimesis—the repeated reference to the imitation of fragments in Latin, Spanish, sounding like animals—does not imply either the corruption of "an originary authenticity" through a badly imitated copy or the hybridization of acoustemes coming from different origins even if the result is what we consider a composite hybrid. Rather, acoustic mimesis is the empirical mechanism through which the acoustic as variation takes form. No "originary essence" is being altered and no hybridity is taking place if the "work" as a concept does not exist. Here culture is understood more "as an on-going act of creation" rather than "the distillation of a set of abstract ideals" (Guss 1989, 4). Culture "consists in the way people draw analogies between different domains of their worlds" (Strathern 1992, 47). This means that the idea of music as "humanly organized sound" (Blacking 1973) is not valid as an ontological basis to distinguish between nonmusical and musical (or linguistic, for that matter) sounds.

Consequently, the problem in encompassing different entities through acoustic ecology is not one of "finding" that other beings "have music" in an anthropocentric extension of our own notion of music. If what acoustic ecology searches for is extending the anthropocentric distinction between sound and music in order to "save" the "environment," the implication of perspectivism would be, to the contrary, to recognize the inherent multiplicity in all sounds and the way they bring together "multiply intertwined materialities" (Feld 2012, 172).

What the bogas would be doing in envoicing such multiplicity is to invoke the transformational potential of becoming that all envoicement entails. It has been said that the exchange of pronouns between beings or parts of the body is a method to name "a transpersonal existence" (Sahlins 2013). If it is so with pronouns, it is even more so with the sonority of animal voices, as vocalizing them implies giving presence to different parts of that transpersonal being and/or to the mutuality established between beings. That is why many musics of the world have no problem in incorporating what we consider "noises," sounds or voices of nonhuman beings.

Here storytelling, songs, and myths do not exist as discrete, isolated entities but seem to be everywhere (Guss 1989; Oliveira Montardo 2009), which could explain the historical difficulty in chronicles of distinguishing between speech, narration, and song and the layering of different symbolic practices despite the identification of clear ceremonial cycles, moments of music-making, and genres. The making of songs, narratives, and discourses, like that of other crafted objects involves an exegetical activity of correlating their symbolic meaning to ritual knowledge and mythical principles and to the powers of transformation that the enactment of singing or narrating invokes (Seeger 1987; Guss 1989; Feld 1996; Montardo 2009). Menezes-Bastos, referring to the musics of lowland South America, speaks of "music as a pivot-system that intermediates, in ritual, the universes of verbal arts (poetry, myth), in relation to plastic and visual expressions . . . and choreologic ones (dance, theater) (2007, 296).

Such musical praxis for the creation of new forms of vocalization that emerged from the encounter between different entities is conceived less as something that originates in an autonomous work and more as an ongoing process of transformation. In this world, "nothing is created, all is appropriated (Sahlins 2013, 57). This would correspond to an understanding of genre as distributive and multiple. Paraphrasing Manuel De Landa, the resultant vocal fragment is more important because of the types of relationality it implies rather than the intrinsic unity that the different parts may have. Such resultant vocalizations are "characterized by relations of exteriority" in such a way that a component part of an assemblage may be detached from it and plugged into a different set of interactions (2006, 10). This takes place in the passage between the ear and the new utterance. In such a world, the basic musical praxis of creating acoustic forms from different acoustic fragments is considered less something that proceeds from "an always already being" (Sahlins 2008, 107), an "oeuvre," rather than a process of transformation understood as a movement between states of multiplicity and unity that can either be expressed as a singularity (a particular "genre") or as a collectivity (a symbolic practice that is not bounded and isolated but simultaneously present in many forms).

This would correspond to a conception of genre as distributive and multiple, as emphasizing relations of exteriority and exchange rather than unity. Specific genres in every culture, of course, are clearly recognizable and distinguishable to the ear, but they are not necessarily defined, as a general concept, by the same principles of boundedness and differential identification that accrue in the idea of representative originality of a particular musical or textual entity that characterizes Western epistemologies of genre and the idea of musical work.

Strictly speaking, then, the bogas' practice of taking sounds from multiple sources does not mean they are enacting some form of musical hybridity. Listening is not solely an exchange practice because it is the locus where acoustic perception is inscribed, where sound trades ears, so to speak, but because it is *transduced* into different types of understandings of musical exchange. The bogas redistribute their acoustic perceptions into vocalizations that use material from different sources and that do not conform to any of the existing genres. This happened again and again in the colonial encounter. A foundational trope for folkloristics and ethnomusicology is the idea that these musics needed to be documented since they were dying not only because the population was being killed, but because the musics they sang were rapidly mixing, through the readiness of natives to copy the music of the colonizers, that is, through the refusal of an epistemology of purification that seeks to separate musical practices discretely into categories of genres that represent people. What we have here instead is that musical invention is understood less as the result of "changing" a genre through transforming its conventional formal elements understood as recurring traits, but rather as an ongoing act of acoustic design through appropriation, which involves the understanding of musical form as a multiplicity capable of constantly incorporating different acoustemes.

From the above, one could conclude, paraphrasing Sahlins, that mimesis as representation is actually quite an exceptional interpretation of the "making and doing" with sounds that we call music. Using Foucauldian terms, the transformation of human beings into subjects or objects of knowledge by the Europeans or Creoles does not mean that the bogas cease to be subjects to themselves or to practice other forms of constructing knowledge. If the mimesis of the bogas undermines the politics of purification as categorization by making whatever form of Spanish intonations to the Virgin and other linguistic fragments their own, thereby upsetting the aesthetic measurements to which such gesticulations and incantations needed to be attuned, then the mimesis of the nineteenth-century European and elite Creole travelers does the opposite: it purifies to establish a developmental history of acoustic materials that highlights discrete entities (pitches and genres) and bounded, unitary wholes as representative of a particular type of being, the racialized, animalized boga.

But addressing the vocalization of the bogas involves exploring not only the conception of music they might have but questioning the distinction between music and sounds, culture and nature. So, I would like to close this chapter by returning to the questions regarding the understanding of sounds of nature that emerged from listening to American nonhuman entities. I now return to Humboldt's "audile techniques" (Sterne 2003) in order to explore his ecology of acoustics.

Humboldt's writing is characterized by a literary style that combines didactic and entertainment purposes, an exceptional sensorial acuity in his description of the environment, and the search for scientific precision of one who tries to find the general laws of the cosmos in the accurate measurement of the facts of nature. Such "echomimetic writing" (Morton 2007) has led contemporary sound installation artists to hail Humboldt as a "forerunner of the acoustic ecology movement, which endeavors to preserve and record the natural sounds of environments" (Velasco 2000, 24).[24] In this section I explore the role of Humboldt's auditory techniques in developing the politics of acoustic ecology that underlies his understanding of nature.

One of the crucial elements of Humboldt's use of auditory techniques is the difference between the way in which he listens to the entities of nature and the commoners who either serve as his guides or whom he encounters in his travels. On repeated occasions, Humboldt discards the questions and conversations of commoners as either impertinent, false, or mistaken in a type of writing that ostensibly racializes his interlocutors. For example, upon arriving in the village of Zapote in the Colombian Caribbean he wrote "the zambos of the Sinú River wearied us with idle questions respecting the purpose of our voyage, our books, and the use of our instruments: they regarded us with mistrust; and to escape from their importunate curiosity, we went to herborize in the forest, although it rained" (Humboldt [1853] 1971, 3:207). Here we find, in narrative form, what later becomes a full-fledged theory of appreciation of natural beauty: nature appears as the *refuge* of the unwanted exchanges with other human beings.

But he certainly listens to his subalterns' knowledge about plants and how they use them. Going farther inland, on the same trip, Humboldt and his companions again found "men of color" at work: "We saluted politely the group of men of colour, who were employed in drawing off into large calabashes, or fruits of the *cresentia cujete*, the palm-tree wine from the trunks of felled trees. We asked them to explain to us this operation, which we had already seen practiced in the missions of the Cataracts (Humboldt [1853] 1971, 3:210). We do not know whether the "men of color" found Humboldt's questions impertinent, but they certainly answered him, since we find immediately after this passage the description they gave him about how palm wine was made.

The zambos or free men of all colors cease to be subjects of speech yet become objects of knowledge in this particular acoustic articulation, as Humboldt becomes the cartographer not only of his own writing (Pérez 2004) but also of his hearing. This erasure of popular sources of scientific knowledge was a common methodology among European and Creole scientists of the period.[25] If we understand "social life as discursively constituted, produced and reproduced in situated acts of speaking . . . linked by interdiscursive ties to other situations, other acts, other utterances" (Bauman 2004, 2), then these are not mere conversations. They are the site of articulation of audile techniques and the traffic of knowledge that constituted the colonial encounter.

Humboldt is a figure who is controversially located at the crossroads of a long debate regarding the relation between Eurocentrism and Americanism (Nieto Olarte 2010). Unlike his central European contemporaries, Humboldt found in the Hispanic American chronicles and documents of the sixteenth and seventeenth centuries a reliable source of information (Cañizares-Esguerra 2001). He used such literature, his personal knowledge gained through travel, as well as what he learned from scientists and commoners in South America, to overturn a Eurocentric colonial discussion of America as a young and immature "new" continent characterized by a backward, degenerate nature (Cañizares-Esguerra 2001, 2006; Nieto Olarte 2007, 2010). Yet recent postcolonial readings of his work have discussed the way he appropriated without acknowledging, crucial ideas from Creole botanists (Cañizares-Esguerra 2006; Nieto Olarte 2007) and indigenous peoples (Serje de la Ossa 2005). He thus occupies a contradictory position in the history of the Americas.[26] His use of audile techniques reveals the nature of some of these tensions by highlighting the way several intellectual trajectories of thought coalesce in Humboldtian natural and civil history: philological diffusionism as a response to Enlightenment, enlightened vitalism as a response to a mechanicist understanding of the universe (Reill 2005), a bourgeois German geography that mixes the aesthetic, the literary, and the scientific in the mapping of the universe (Farinelli 2009; Minca 2007), and a literary style of scientific writing that, by the end of his life, was considered outdated due to the increasing distinction between the humanities and the sciences (Rupke 2008). When exploring the politics of his audile techniques, these different elements come together.

In Humboldtian physical history of the universe, the knowledge of primitive peoples rests on "primitive intuitions" (1858, 23) derived from the awe produced by nature: "In the earliest stages of civilization, the grand and imposing spectacle presented to the minds of the inhabitants of the tropics could only awaken feelings of astonishment and awe. It might be supposed, as we have said, that the periodical return of the same phenomena, and the uniform manner in which they arrange themselves in successive groups, would have enabled man more readily to attain to a knowledge of the laws of nature; but as far as tradition and history guide us, we do not find that any application was made of the advantages presented by these favored regions" (Humboldt 1858, 36). Thus it is to "the inhabitants of a small region of the temperate zone that the rest of mankind owes the earliest revelation of the intimate and rational acquaintance with the forces governing the physical world" (Humboldt 1858, 36). In Humboldt's natural philosophy then, the feelings of grandiosity produced by the overwhelming American nature need to be linked to the European rationalist discernment of its laws.

In order to trace "the enjoyment derived from the exercise of thought" he turns to "the earliest dawnings of a Philosophy of Nature or of the ancient doctrine of the Cosmos" (36) and finds "even among the most savage nations ..., a certain vague, terror-stricken sense of the all-powerful unity of natural forces, and of the existence of an invisible, spiritual essence manifested in these forces" (Humboldt 1858, 37). Humboldt rightfully perceives the connection between cosmology and lived, experienced place, so crucial to many peoples. However, he seeks to transcend it by separating the sensorial. So he seeks to "trace the revelation of a bond of union, linking together the visible world and the higher spiritual world which *escapes the grasp of the senses* (emphasis mine). The two become unconsciously blended together, developing in the mind of man, as a simple product of ideal conception and independently of the aid of observation, the first product of a *Philosophy of Nature*" (Humboldt 1858, 37).

The primitives, however, do not separate "the world of ideas from that of sensations" (37) whereas in a man "that has passed through the different gradations of intellectual development" one finds "the free enjoyment of reflection" (37). In the primitives, instead, thought is derived from their "predilection for symbols" that leads to "conjuncture and dogmatism" rather than to facts derived from observation (37). So whatever brings enjoyment to thought is the capacity of overcoming what the senses cannot grasp by studying the laws of the cosmos as a means to spiritual transcendence. Enthrallment with the senses, including observation, needs to be transformed in order to enable such enjoyment, something the primitives cannot do.

The purpose of Humboldt's natural philosophy "is to combat those errors which derive their source from a vicious empiricism and from imperfect inductions" (Humboldt 1858, 38). Thus the incapacity of primitive people or uneducated ones to discern laws of nature through reason leads to the "assemblage of imperfect dogmas, bequeathed from one age to another—the physical philosophy, which is composed of popular prejudices" (38). This "is not only injurious because it perpetuates error with the obstinacy engendered by the evidence of ill-observed facts, but also because it hinders the mind from attaining to higher views of nature" (38). This produces, instead of the utter enjoyment of reason, preposterous popular fears (43). Thus, the perception of the knowledge of "savages" in the voyages of nineteenth-century planetary science serves to interpret the unreason of past ages and the contemporary empiricism of popular culture in the metropolis, giving rise to the global history of "the popular." In Humboldt's developmental system of history, the monumental architecture of past ages was the sign of great civilization, while the dispersed habitation of the rainforest was a sign of primitivism, of an intuitive and sensorial relation to the world rather than an intellectual one. But in Western developed peoples, such sensorial intensity to intellectual deduction and the joy it brings could be challenged into transcendence.

Humboldt is obsessed by the relations between different elements of the cosmos, "the inextricable network of organisms" (1858, 41), the "knowledge of the laws of nature" (42) and the need to "consider each organism as a part of the entire creation, and to recognize in the plant or the animal not merely an isolated species, but a form linked in the chain of being to other forms either living or extinct" (42). This capacity for accessing the occult in physical phenomena is central to Humboldt's definition of nature as an "occult but permanent connection, this periodical recurrence in the progressive development of forms, phenomena, and events, which constitute *nature*, obedient to the first impulse imparted to it" (50). Such are also the doors to intellectual enjoyment, a preamble to the way both sciences and the "fine arts" are conjoined in the pursuit of noble purposes: "As in nobler spheres of thought and sentiment, in philosophy, poetry and the fine arts, the object at which we aim ought to be an inward one—an ennoblement of the intellect" (53). It is through the understanding of general laws and principles that "physical studies may be made subservient to the progress of industry, which is a conquest of mind over matter. By a happy connection of causes and effects, we often see the useful linked to the beautiful or the exalted" (54). Nations also function as one organism to be molded by such actions, with inaction, according to Goethe, cited by Humboldt, being their worst curse: "Communion with nature awakens within us perceptive faculties that had long lain dormant: and we thus comprehend at a single glance the influence exercised by physical discoveries on the enlargement of the sphere of intellect, and perceive how a judicious application of mechanics, chemistry and other sciences may be made conducive to national prosperity" (52). In this fragment, enlightened vitalism's emphasis on the occult is linked to Humboldtian philological geopolitical emphasis on the differences between nations explored in the earlier part of this chapter. In what follows, I explore how this relation between science, transcendence, the senses, the occult, and the significance of the difference of interpretation of different peoples, appears in his practices of listening to nature.

Part of Humboldt's acoustic perception became a general law, known as Humboldt's acoustic effect, which describes the increase of the volume of a sound by night and in lower temperatures. As analyzed in a lecture to the Academy of Sciences in Paris in March 1820:

> The noise is three times as loud by night as by day, and gives an inexpressible charm to these solitary scenes. What can be the cause of this increased intensity of sound in a desert where nothing seems to interrupt the silence of nature? The velocity of the propagation of sound, far from augmenting, decreases with the lowering of the temperature. The intensity diminishes in air agitated by a wind, which is contrary to the direction of the sound; it diminishes also by dilatation of the air, and is weaker in the higher than in the lower regions of the atmosphere where the number of particles of air in motion is greater in the same radius. (Cited by Velasco 2000, 23)

Let us contrast the description of this particular law with his writing on how he came to hear such a law. It appears in his *Views of Nature: Or Contemplations on the Sublime Phenomena of Creation, with Scientific Illustrations* (1850). It is one of the scenes of his trip down the Orinoco, the Casiquiare, the Rio Negro, and the Apure. The passage is long but well worth quoting, a classic of Humboldt's detailed, attentive listening:

> Below the mission of Santa Barbara de Arichuna we passed the night as usual in the open air, on a sandy flat, on the bank of the Apure, skirted by the impenetrable forest. . . . Deep stillness prevailed, only broken at intervals, by the blowing of the fresh-water dolphins . . . After eleven o'clock, such a noise began in the contiguous forest, that for the remainder of the night, all speech was impossible. The wild cries of animals rung through the woods. Among the many voices which resounded together, the Indians could only recognize those which, after short pauses, were heard singly. There was the monotonous, plaintive cry of the Aluates (howling monkeys), the whining, flute-like notes of the small sapajous, the grunting murmur of the striped, nocturnal ape (*Nyctipithecus trivirgatus*, which I was the first one to describe), the fitful roar of the great tiger, the Cuguar or maneless American lion, the peccary, the sloth, and a host of parrots, parraquas (*Ortalides*), and other pheasant-like

birds. Whenever the tigers approached the edge of the forest, our dog, who before had barked incessantly, came howling to seek protection under the hammocks. Sometimes the cry of the tiger resounded from the branches of a tree, and was then always accompanied by the plaintive piping tone of the apes, who were endeavouring to escape from the unwonted pursuit.

If one asks the Indians why such a continuous noise is heard on certain nights, they answer, with a smile, that the "animals are rejoicing in the beautiful moonlight, and celebrating the return of the full moon." To me the scene appeared rather to be owing to an accidental, long-continued and gradually increasing conflict among the animals. Thus, for instance, the jaguar will pursue the peccaries and the tapirs, which, densely crowded together, burst through the barrier of tree-like shrubs which opposes their flight. Terrified at the confusion, the monkeys at the top of the trees join their cries with those of larger animals. This arouses the tribes of birds who build their nests in communities, and suddenly the whole animal world is in a state of commotion. Further experience taught us that it was by no means always the festival of moonlight that disturbed the stillness of the forest; for we observed that the voices were loudest during violent storms of rain, or when the thunder echoed or the lightning flashed through the depths of the woods . . . A singular contrast to the scenes I have described, and which I had repeated opportunities of witnessing, is presented by the stillness which reigns within the tropics at noontide of a day unusually sultry. I borrow from the same journal the description of a scene at the Narrows of Baraguan (in the Orinoco). . . . The larger animals at such times take refuge in the deep recesses of the forest, the birds nestle beneath the foliage of the trees, or in the clefts of the rocks; but if in this apparent stillness of nature we listen closely to the faintest tones, we detect, a dull, muffled sound, a buzzing and humming of insects close to the earth in the lower strata of the atmosphere. Everything proclaims a world of active, organic forces. In every shrub, in the cracked bark of trees, in the perforated ground inhabited by hymenopterous insects, life is everywhere audibly manifest. It is one of the many voices of nature revealed to the pious and susceptible spirit of man. (Humboldt 1850, 198–200)

Audition is the corollary sense to observation for understanding the "occult" connection between different phenomena of sound in nature in order to produce an acoustic law that enables the transcendence of the overwhelming sensorial impressions that produced it. But here Humboldt's transcendence and his sensorial acuity extend beyond the instrumental reason with which he links the fine arts to science (in order to ennoble the spirit). Here, the "many voices of nature" reveal rather his understanding of "man" through acoustic sensation. The loudness and noise of the forest signals "the increasing conflict among animals" while the "stillness" of the "faintest tones" of nature yield the "many voices of nature revealed to the pious and susceptible spirit of man." Thus quietness and the silence of nature provide a spiritual allure that translates as the acoustic equivalent of the miracle of the occult, a "haven of interiority and self-identity" (Viveiros de Castro 1992, 3), an issue central to the grandiose mythological character of German bourgeois geography (Farinelli 2009). Such silence emerges in contrast with the conflictive loudness and noise of the animals. It is quite telling that the reflection on silence of nature as a form of refuge comes immediately after the condemnation of cannibalism of the tribes of the Casiquiare and Rio Negro region, extrapolating from such practice a "sinister view of human nature" (Sahlins 2008, 3):

> Thus does man, everywhere alike, on the lowest scale of brutish debasement, and on the false glitter of his higher culture, perpetually create for himself a life of care. And thus too, the traveler, wandering over the wide world by sea and land, and the historian who searches the records of bygone ages, are everywhere met by the unvarying and melancholy spectacle of man opposed to man.
>
> He, therefore, who amid the discordant strife of nations, would seek intellectual repose, turns with delight to contemplate the silent life of plants, and to study the hidden forces of nature in her sacred sanctuaries; or yielding to that inherent impulse, which for thousands of years has glowed in the breast of man, directs his mind, by a mysterious presentiment of his destiny, towards the celestial orbs, which, in undisturbed harmony, pursue their ancient and eternal course. (Humboldt 1850, 21)

This passage echoes the one on the conflictive noise of the animals by contrasting "the lowest scale of brutish discernment" with the repose provided by the silent life of plants. The many voices of nature then yield a contrast between loudness/noise and silence as signaling a metaphysical difference between humans and nonhumans. In Humboldt's ecology of acoustics the transcendence of the sensorial in the joy of thought that leads to the ennoblement of the spirit is accompanied by a separation between human and animal nature. Thus, while in Humboldt's vitalism the links between organisms are fundamental, such links are also used to differentiate between different peoples and between people and other entities of nature, something made evident by his audile techniques for listening to nature.

As Humboldt was busy losing his soul into the silent nature and drawing acoustic laws from the noisy animals at night, the indigenous peoples who guided him were thinking that animals were thinking that they were "rejoicing in the beautiful moonlight and celebrating its return"; that is, engaging in the type of nocturnal singing that is characteristic of peoples in this region, reminding us of a different relation between silence and noise. In this history of aurality at the crossroads of late colonial encounters we find different understandings of the relation between sounds, music, people, and places. Aurality is central to the constitution of ideas about Latin American nature and culture and, in turn, ideas about nature and culture are central to the constitution of understanding of the relation or distinction between sound and music. These histories then also imply different ecologies of acoustics.

In all cases explored in this chapter, silence and noise seem to be the acoustic parameters that serve as the auditory entry points into questions about the nature of life. In his analysis of the role of sounds in myths about the limited duration of human life, Lévi-Strauss formulated the following question: "May we not suppose that in these three cases [the three myths he analyzes] the nature of life on earth, which is—by its limited duration—a kind of mediatization of the contrast between existence and non-existence, is being thought of as a function of man's inability to define himself, unambiguously, in relation to silence and noise?" ([1964] 1983, 149). The layering of different metaphysics of sound has been a central dimension of Latin American history, couched in the way the acoustic lies in the different understandings of the given and the made, of convention and invention, and in the impossibility of establishing an unambiguous relation with noise and silence when faced with trying to define the nature of life on earth.

History is a songbook for anyone who would listen to it.
—R. Murray Schaffer

José María Vergara y Vergara's *Historia de la literatura en Nueva Granada desde la conquista hasta la independencia (1538–1820)* (History of Literature in New Granada from Conquest to Independence, 1538–1820) was first published in Bogotá in 1867. The topic of popular song, placed strategically at the beginning and end of the book, plays a prominent role in his historiography since it sets both the foundational logic for such a history and the programmatic project for its continuation. By the beginning of the twentieth century, this book had become canonical for studies of popular song or "popular poetry" as it was called then.

Its structure and theoretical purpose were fundamentally reproduced and augmented in Gustavo Otero Muñoz's literary history (1928). After complaining about the lack of "systematic studies" of folklore in Colombia, Otero admits that there have been "occasional folklorists that have compiled in literary or scientific works of diverse character, news belonging to such topics" (1928, 242). Among them, he mentions Jorge Isaacs and his novel *María*, which "gives us the general tone of the dialogue and in the description of nature, a faithful picture of the speech (*el habla*) and the land of Cauca,[1] with approximately 200 provincialisms, which will live as long as Isaacs's creation" (244). He also mentions Candelario Obeso, author of *Cantos populares de mi tierra* (Popular songs of my land), a beautiful collection of poetry written in the uncultured language of the *boga* from the Magdalena" (244). Candelario Obeso and Jorge Isaacs are the authors of the only two known nineteenth-century popular song collections in Colombia. This chapter is about the ways that song was inscribed into "the scientific and the literary" by José María Vergara y Vergara, Candelario Obeso, and Jorge Isaacs through similar techniques of orthographic manipulation of sound, used, however, with very different political ends and aesthetic intentions.

At the most obvious level such a difference lies in the locus of enunciation from which the intellectual projects of these authors were articulated in a country torn apart by the many civil wars that characterized this period.[2] José María Vergara y Vergara (1831–1872) was the founder and director of several political and literary journals published in Bogotá, the first director of the Colombian Academy of Letters created in 1871 (the first one in the Americas), founder and director of an informal but highly influential literary circle called El Mosaico (concerned primarily with "local" culture), and a key figure in the production of an ideology of Hispanic Catholic conservatism that came to dominate the country by the 1880s (Von der Walde 2007). Jorge Isaacs (1837–1895), writer, soldier, diplomat, ethnographer, and educator, was the son of a Jewish immigrant from Jamaica and a Catholic Italian-Catalan woman who had settled in the city of Quibdó on the Pacific Coast, then part of the Cauca state, a predominantly Afro-Colombian region. He is the author of *María*, one of the most famous novels of the nineteenth century in Latin America. He also wrote political treatises, theater pieces, two unfinished novels, several poems, and one of the earliest Colombian ethnographies, *Estudio sobre las tribus indígenas del Magdalena*, which we will explore in detail in chapter 4. Although, in his early years, he was highly valued by the circle of El Mosaico, who hailed his novel as one of the greatest depictions of local culture, he ended his life highly despised by them due to his overt defiance, as a soldier, educator, and writer, of their Hispanic Catholic conservatism.

Candelario Obeso (1849–1884) was a philologist, translator, poet, diplomat, and man of arms who has been described as Colombia's first major writer of African descent (Caraballo 1943; Prescott 1985). He also wrote grammars, theater pieces, texts of applied linguistics for the teaching of Italian, English, and French in Colombia, political essays, and collections of poetry beyond those that comprise his famous songbook. He was born in the Caribbean city of Mompox, but spent his later years in Bogotá. For his role in the battle of Garrapatas, Obeso earned an honorable promotion to *sargento mayor* (major sergeant). The Battle of Garrapatas was instrumental in the transition from Radical Liberalism, which dominated the country between 1863 and 1878, to the Conservative government's alliance with a faction of Independent Liberalism that eventually led to the highly centralist and conservative period called the Regeneration (1878–1903) and the political dominance of the Conservative party in the early years of the twentieth century (1886–1930) (Ortiz

Mesa 2005; Sanders 2004).[3] Candelario Obeso and Jorge Isaacs, among other notable writers and philologists of the period, often changed their lettered uniforms for those of soldiers and fought battles in the name of the Radical Liberals, who were then in power, against their Conservative brethren.

The versatility of these men "was not rare in Colombian public life" (Deas [1992] 2006, 27). And the political struggle they enacted was not only military. It also included the use of grammar, philology, and attendant activities such as the collection of local songs and "provincialisms." By the end of the century, and in the hand of the Catholic Conservatives, philology had become a discipline cultivated by grammarian-presidents who prescribed linguistic uniformity as a key element of unification of the nation and grammatical correctness as an articulator of governmental rhetoric and power (Rodríguez García 2010; Sierra Mejía 2002; Von der Walde 1997). The term *grammarian*, in Colombia, usually refers to philological presidents of the end of the nineteenth century associated with the rise to power of the Conservative party.[4] But in the mid to late nineteenth century, the cultivation of philological endeavors was mediated by scholars of many different political persuasions and with a wide variety of interests. The differences between Vergara y Vergara's, Isaacs's and Obeso's politics of acoustic inscription are highly significant in that they reveal how the problem of heterogeneity of populations as one that defies a clear political relation between an imagined community and literary texts or folkloric expressive genres that are supposed to embody and contain it. In this chapter I explore why popular song was inscribed into the literary rather than the musicological domain.

The idea of popular song as a particular materialization of the relation between language and music also brings to the foreground a "nonlinear history" (De Landa 1997), one that seems to be more related to the peculiarities of song as an entity rather than to its discursive genealogy. In histories of folklore, one is generally taught that the geopolitical, social, and temporal components that characterize the history of the popular are primarily grounded in an eighteenth-century central European Romanticism that challenged universalistic, rationalistic, and progress-oriented Enlightenment theories by tying folk song to place, to affect, to the spontaneous, and to "the people."[5] As we shall trace in this chapter, in the Ibero-American world such genealogy is more related to a Hispanic nineteenth-century revival of Renaissance poetics or to Ibero-American struggles over diversity than to central European Romanticism. But the idea that songs are capable of enacting the relationship between place, personhood, affect, and time in particularly intense ways is also present in places where its epistemological articulation has nothing to do with European Romanticism or Hispanic nineteenth-century nationalism and their expansionism.[6] If the search for a specific intellectual genealogy leads us to trace concrete networks of interaction and practices of inscription that help us understand the epistemological production of popular song, it does not necessarily answer the question of why song is repeatedly seen, in many different places and in different historical moments, as a field of force capable of enacting translations between space, time, affect, and different beings.

Song's materiality enables the production of such hybrids under very different historical and political circumstances due to several factors. First, one of the central properties of songs is their capacity to incorporate great mobility. Songs are easily transduced (inscribed from one medium to another and back), translated (in the specific sense of changing the language of the song), transported (taken from one place to another through different media), adapted (malleably changing form as when a song sung by a single singer is taken up by a whole audience), arranged (transformed musically without losing their identifiable shape), and transferred from one person or group of people to another. Second, song, more than any other genre of orality, brings together the heightened orality of poetry with the aurality of music (Fox 2004). Third, due to its malleability, song can easily become a part of another artifact such as a songbook, a movie, or a CD, thereby having the potential to metamorphose and exist as part of another form. It can thus be recognized as being "the same" yet different. Fourth, its capacity of malleability and intermediality make it potentially able to adapt and be adopted across temporal changes, an entity that constitutes repetition, recurrence, and difference across time and across its many material supports. Because of this, it can easily be amenable to discourses on the temporal, be they those of conservation and stasis or of change and futurism. Finally, because of all of the above, song seems particularly structured to pull the strings of affect (Gray 2013). Thus, as I explore the genealogies that help us trace the ideologies that surround the emergence of song as a national preoccupation in the initial years of the postcolony, I am simultaneously looking at a history of usage and transformation of song's "materials" (Ingold 2011, 26), that is, the "stuff" songs are made of and how they change. The three scholars explored in this chapter highlight different dimensions of song's materials and potentialities in their modes of inscribing song into writing.

The question of music and language as different fields of signification that intersect in song (Fox 2004), and its consequent ambivalence (Samuels 2004), has often been posed as a question of how songs "mean," answered through specific socio-musical and sociolinguistic analyses that explore the heightened arbitrariness of an acoustic sign that simultaneously enacts the indexical and iconic capacities of language and music (Fox 2004; Meintjes 2005; Samuels 2004; Weidman 2006). Song brings to the foreground the entangled histories of the musicological, the literary, and linguistic as mutually constituted domains of knowledge (Feld and Fox 1994). As I begin to discuss in this chapter, the differentiation in central Europe in the nineteenth century between the philological and anthropological sciences, on the one hand, and the musicological disciplines, on the other, emerges from the global history of comparativism. Such a history has to do with the way the comparativist moment of the late nineteenth century set up the distinction between "orality" (or the spoken/written in verbal arts) and (absolute) music through the emergence of the work concept. But song's ambivalence questions this historical disciplinary distinction as applicable to disciplinary domains everywhere and helps us understand why in the study of popular musics in Latin America and the Caribbean—particularly of verbal expressive genres—this distinction did not fully coalesce. Thus, in this chapter we explore a different sonic history of comparativism, one in which the musicological, linguistic, and anthropological are deeply entangled. Central to such a history is not only the politics of the site of enunciation of particular scholars but also this history of ambiguity of song as a form and the centrality of mishearing in producing its ambivalent disciplinary inscription.

Song brings to the foreground "the intolerance of pure phenomena" that often characterizes popular expressions (Connor 2009, 3). In doing so, it highlights a particular history of mishearing, evident, for example, in the way that the different collectors or song theorists described in this chapter transfer their audile techniques into writing. According to Connor (2009) "mishearings often attach themselves to popular or traditional forms of utterance" where what is significant is the creative potentiality of (mis)understanding contained in the "earslip" (Connor 2009, 1). However, "seeing slips of the ear as simply the auditory complement of slips of the tongue mistakes their programmatic nature and function. Misspeakings are the disorderings of sense by nonsense; mishearings are the wrenchings of nonsense into sense" (Connor 2009, 3–4). Such mishearings and their wrenching into sense are central to the politics of inscription. These inscriptions that bear the trace of the "slip of the ear" in turn speak of the relation between auditory sensation and auditory perception. Auditory sensation "may be viewed as processing the physical attributes of sound: frequency, level and time. Auditory perception may be the additional processing of those attributes that allow an organism to deal with the sources that produced the sound" (Yost 2008, 1). These attributes are: locating the sound source and the perceiving organism in space (as in how a person manages to make decisions about walking or driving by virtue of hearing a car coming); its temporal perception (the issue of acoustic memory since sound is always arriving and evanescing by bits as we hear it); and the segmentation of simultaneous sounds into differentially perceived acoustic elements (that which helps us hear the voice of a friend in the midst of a noisy party) (Yost 2008). All of this then is related to the role of acoustic feedback in the process of auditory inscription: "the ear subtly and actively connives to make what it takes to be sense out of what it hears, by lifting signals clear from noise, or recoding noise as signal. In other words, listening is full of replay, relay and feedback, the ear monitoring or listening in on, and out for, its own operations. Perhaps, in this sense, all hearing is mishearing, and a kind of deterrence of sound" (Connor 2009, 9). Understanding such a relation complicates the idea of "site of enunciation" as a transparent sociological attribute. The history of production of such a heightened acoustic relation between the ear and the tongue, between mishearing and the politics of song inscription, unfolds through the feedback between sounds, silences, graphemes, geographies, politics, religion, practices of racialization, and peoples. In what follows, I explore the different modes of inscribing song into writing in nineteenth-century New Granada as part of such global history.

Song in Vergara y Vergara's Literary History

ON SONG AS THE ORIGIN OF LITERATURE

Vergara y Vergara's book begins with a reference to fifteenth-century Castilian song and prose poetics and ends with a concluding chapter on "popular poetry" in which he describes songs from several regions of Colombia. Song is also mentioned when he hypothesizes about the possible expressive realms of the Muisca Indians, who lived in the region where the Spanish conquerors founded Bogotá. These three moments figure song into "national time" (Lomnitz 2005) as marked by two beginnings and one ending: the conquest and the beginning of colonial times as the beginning of New Granadian literary history, the end of indigenous time marked by the disappearance of indigenous song and the death in battle of indigenous heroes in times of conquest, and the beginning of Republican times (at the end of the book, Vergara y Vergara's present) as that of the search for national popular song. Yet all these beginnings and endings are interdiscursive, framed by texts that point to texts within texts, as if Vergara y Vergara were locating himself within a lineage of potential endless repetition, replay, and relay. He casts literary history as a heroic enterprise of discovery and collection in "a vast region from which I bring samples" just like the early soldiers of conquest brought samples of the newly discovered lands to Spain (1867, xxiv). The history of literature is, at least in this particular sense, not that different from the scientific sampling characteristic of natural history.

The book opens by setting the stage for such collecting of samples by describing "the state of literature in the peninsula at the beginning of the sixteenth century," the time of "a literary Spain so often confused with a warrior Spain" (1867, 1). The beginning of Spain's literary glory coincides with the reconquest of Spain from the Arabs and the conquest of New Granada: "literature was in the last day of its infancy, and was about to fully enter its healthy and vigorous youth when the armies destined to conquer New Granada were readying themselves in the coasts of Santa Marta" (2). Such embattled literary affairs rested on the interrelationship between poetry and language:

> The affection for the literary, a sentiment that had arisen under the Kingdom of Juan II, stimulated by the King himself and his assistant, who also wrote *coplas*, was being spread throughout those untiring warriors; and if it had a period of decline which we do not wish to remember, under the Kingdom of Alfonso, it had a strong return under the Kingdom of Isabel and Fernando. When these kings turned over their vast and flourishing empire to Charles V, it was found that language and poetry had become marvelously polished in the silence of peace and in the shadows of the laurels of glory. (Vergara y Vergara 1867, 2–3)

Vergara y Vergara establishes a direct link between song craft, the birth of the vernacular, the wars of reconquest in Spain and the conquest of New Granada, thus recasting the practice of Spanish poetics from that of a national enterprise to a colonizing, imperial one. Sheldon Pollock states that the "vernacular literary cultures were initiated by the conscious decisions of writers to replace the boundaries of their cultural universe by renouncing the larger world [of Latin cosmopolitanism] for the smaller place, and they did so in full awareness of the significance of their decision" (Pollock 2000, 592). But in the case of Spanish, the vernacular also became, scarcely two centuries after its consolidation as an administrative national language, the imperial language of conquest in the Americas, considered central to the spread not only of the Spanish empire but also of Christianity (Mignolo 1995; Lomnitz 2001). That Spanish became identified as the language of Christianity for the Ibero-American world was in large measure predicated on the simultaneous deployment of religious unification through the expulsion or annihilation of so-called idolatry in Spain and in the Americas (Bernard and Gruzinski 1988), the adoption of Spanish as an administrative and literary language in Spain, and the expansion of the Ibero-American empire. Such a history then is not solely the history of language in the colonies but also a crucial piece in the history of the creation of the idea of "religion" (Asad 1993) and of "the occult" as a fall into secrecy, a silencing of certain practices, a spectrality produced as a by-product of the repressions of the Inquisition (Thacker 2011) and of conquest.

The mention of Juan II, "the king and his assistant who made coplas" and who stimulated their composition, is a reference to the literary glories of the kingdom of Juan II and, indirectly, to the *Cancioneros* (Songbooks) of the period, and probably and more specifically to the *Cancionero de Baena*. The *Cancionero de Baena* is considered a foundational work in Spanish philology. It is a songbook compiled by Juan Alfonso de Baena (c. 1375–c. 1434) for Juan II of Castile (himself a composer of poems) during a period of intense poetic creativity in Spain that, according to Julian Weiss, was "unmatched by any other European country" (1990, 2). This period is reflected in the preservation of "nearly two hundred manuscripts and over two hundred printed anthologies" of fifteenth-century court poetry at the moment of consolidation of a "vernacular humanism" (1990, 2).[7] By virtue of having been rediscovered and published in the middle of the nineteenth century, the *Cancionero de Baena* "became a starting point for scholarship on fifteenth century poetry" (Beltrán Pepió 1998, 19–20) in Spain and songbooks in general came to represent a central aspect of nineteenth-century Spanish philology.

Such medieval poetry was not simply rediscovered but also reinterpreted in crucial ways. In medieval times, the poetic creativity of the period was characterized by a fluidity between the popular and the cultivated that did not hierarchically subsume one into the other (Funes 2003); by an emphasis on the practices of enunciation of poetry (sung or recited out loud) as that which framed its particular features, rather than a theoretical preoccupation with identifying a clear-cut genre (Gómez-Bravo 1999); and by a "lack of discrete limits" in practices of naming "genres" that functioned more to establish communicative continuities with the audience than to provide systematic classifications for canonization (Higashi 2003). But such social and formal fluidity was overturned in the Spanish nineteenth-century reinterpretation of medieval poetics: "one of the first operations of nineteenth-century literary history was locating the monuments of medieval textuality in the pantheon of the origins of high literature" (Funes 2003, 24). This was done through a "creative (mis)understanding of history" that drew upon "genre conventions and stylistic details" (Luker 2007, 69) of the previous historical period but adapted them to nineteenth-century Spanish poetics. In turn, Vergara y Vergara creatively recasts such a politics to frame New Granadian literary history.[8] Its foundational gesture is less about the Conquest as the original literary moment than about providing a connection between New Granada and Spain by walking on a path that emerges as he delineates it with and through literary forms.

Besides this spatial and temporal relocation of Spanish song, Vergara y Vergara mentions another crucial element that further characterizes his foundational poetics. In his book he distinguishes between "traditional" Renaissance Spanish poetry (influenced by popular song) and "modern" Renaissance poetry, represented by Italianate Petrarchan poetry. This was supposed to be a "cultured" type of poetry that exacerbated the division between uncultured and erudite works, in which the latter were marked by foreign (Italian) influences. According to Vergara y Vergara, the former "more general language, more used, more uncultured and incorrect, was the one brought by our conquerors and the one that was spoken and written for a long time in the New Kingdom [of Granada]" (Vergara y Vergara 1867, 3). The emergence of the popular poetry of *juglares* (minstrels) in Spain was also seen as foundational to the rise of Spanish as a literary language and of Spain as a nation that conquered the Infidels in Pidal's introduction to the *Cancionero de Baena* (Pidal 1851). In an evident parallelism, in Vergara y Vergara's literary history the emergence of low-class, "uncultured" poetry and language is associated with the arrival of a warrior caste of conquerors and with the origins of literature history in New Granada, mentioned earlier. However, and, for Vergara y Vergara, unfortunately, this "uncultured" language did not lead to an appropriate foundational song practice in New Granada:

If Miguel de Espejo, Cristóbal de León, Sebastián García and others that made rhymes in that time, instead of writing erudite poems, and since they had imagination and good taste, had taken the route of writing *romances* (ballads), for which they had a model in those of El Cid; if, instead of celebrating everyday events, such as the publication of a book, they had celebrated wondrous events such as the feats of the conquerors, of the Indians or the beauty of this land, they would have founded a national and rich literature, in which they would have collected all the traditions that were then fresh because the Spanish heroes were then alive and so were the sons of the Chibchas.[9] When the two types of poetry, the erudite and the popular, were struggling in Spain, it was natural that the people would have rebelled against the pedagogy that oppressed them in Madrid; and, that they would have sung here, in liberty and spontaneity, in the midst of the American jungles, the same things that they sang under cover in Spain, having more subject matter here than in Spain for their songs. The colony itself, with its picturesque life, lent itself, and still does, to the *romance* (ballads). (Vergara y Vergara 1867, 18)

Vergara y Vergara makes clear that there is a distinction between popular song, which makes possible a foundational account because of its epic content, and badly copied erudite poetry, which does not serve a foundational purpose, not only because of the failure of the mimetic colonial imagination (it is a bad imitation of Spanish poetics) but also because the lyrics of the songs composed by Spanish soldiers were deemed by Vergara y Vergara as irrelevant for such a purpose.[10]

The epic is a genre for producing the temporality of grand events, one that is capable of covering, in sung narration, "great spans of time" (Benjamin [1955] 1968a, 149). It is the type of song called on to fill the "abyss" between "a circumstance which has been too little noticed" (Benjamin [1955] 1968a, 154) and its abstraction into a foundational event of the nation. But for Vergara y Vergara, the lack of a genre to do so raises the impossibility of bridging such a gap. Interestingly enough, however, Colombia did have a national epic and chronicle of conquest called *Elegías de varones ilustres de Indias*, written in verse by Juan de Castellanos. Vergara y Vergara praises it ("the poet that dealt with things of the Indies merits great praise" (20), but he does not give it the status of a foundational epic, such as he does to *La Araucana*, the Chilean chronicle of conquest, effectively erasing Juan de Castellanos's text as a foundational epic.

Such a failure is related to another one: that of the early chroniclers to preserve the indigenous epic songs that, surely, the valiant Muisca soldiers fighting against the Spanish invaders must have been singing at that moment. After complaining about the lack of chroniclers who properly documented indigenous customs of the Muiscas and expressions of New Granada at the time of the conquest, Vergara y Vergara speculates about the possible nature of such "literature":

It is natural to believe that the Indians had their poets, just as all peoples do. Among the Muiscas, the *mohán* was probably the most inspired, following the footsteps of all men from the infancy of nations, even the most uncultured ones. The men that find some imagination begin by singing to their gods, and later they celebrate the feats of their heroes. Then their own throat itself incites them to song, in the excitement of feasts, combat, or religious celebrations. One can hear their delirious, harmonic, or ardent words, as is always the case when speaking with the soul; and some years later one can hear the rhapsodies repeated by the people. The Chibchas probably also had their religious songs and their warrior hymns, that they surely sang in the monotonous recitation characteristic of the beginnings of song of our barbarian peoples. (Vergara y Vergara 1867, 14)

The speculation on the "monotonous" yet epic recitations of Indians is based on his citation of Lucas Fernández de Piedrahíta (1624–1688), the main New Grenadian chronicler of conquest whose descriptions of "Chibcha" song and dance Vergara y Vergara cites at length. For Vergara y Vergara such a song disappeared during early colonial times because indigenous languages became, through the conquest, a "vanquished tongue," the only "remnants" (*despojos*) (1867, 508) by the nineteenth century being "some provincial words of some indigenous objects" (509). Spanish "took some time in becoming the original language; but in the end, it obtained victory, and became the only, sovereign and dominating language" (508). The erasure of indigenous song then depends on the postulation of the disappearance of indigenous peoples and languages through conquest thus giving Spanish an originary and sovereign status. Song theorization is thus overdetermined by the politics of language.[11]

Despite the heroic geopolitical rhetoric, Vergara y Vergara nevertheless proposes a rather impoverished linguistic sovereignty: at the moment of conquest the soldiers spoke an uncultured Spanish that became the one spoken and written in New Granada for a long time, did not compose the appropriate *romances,* failed to document the epic heroism of indigenous peoples, and such a chronicle as there was (Castellano's *Elegías ilustres de Indias*), was poetically significant but not worthy as a foundational epic. Such a patriotic historiography then is based on two fundamental premises. The first is that the foundational failure he is at pains to demonstrate posits Vergara y Vergara's literary history as the real foundational moment. If others did not collect or compose appropriate samples, he is doing so by inaugurating a "novel regime of marks and meanings" (Anidjar 2008, 16) through his literary history. Second, he does so guided by the same Hispanic Catholic poetic ethos of the original conquerors since he casts his literary history as a religious enterprise to be *sung* in hymnal mode:

My book is nothing more than a long hymn sung to the Church. For this I will offer no excuses. I wanted to write only a literary history. . . . But since what I was searching for, the lettered (*las letras*), I always find in the bosom of the church itself, I had no reason to hide it. I did not need to negate that it is very pleasurable for me to bring together the glories of the church and those of the fatherland. I wish that all my works were in the service of the Catholic cause . . . so, if the reader that takes this book does not like Catholic writings, he should abandon it right from this page. (1867, xxii–xxiii)

Like the Spanish Reconquest, Vergara y Vergara's literary project brings together the church and the nation through language, war, and song. Angel Rama's description of the men of letters in the lettered city as a "priestly caste" ([1984] 1996) emerges here not as a metaphorical insight but as a literal transposition: the creation of a literary canon enacts the musico-religious canonization of its producer. And the relationship between language and Catholicism, the requirements for citizenship in the lettered city in Vergara y Vergara's Hispanic Catholic conservatism, are underscored by the political theology of the hymn. The nineteenth-century civil wars, of which one of the major conflicts was the place of the Catholic Church in the public sphere (González González 1997), became the site of deployment of such epico-religious enterprise.

THE POLITICS OF SONG INSCRIPTION, LITERARY THEORY, WAR

For Philippe Lacoue-Labarthe and Jean-Luc Nancy, the multiple political crises of Germany in the late eighteenth century gave rise to the Romantic movement and to the emergence of a historical period of literary theory that still persists today. It is still thought of "as the critical age par excellence . . . or, in other words, as the 'age' (almost two hundred years, after all) in which literature—or whatever one wishes to call it—devotes itself exclusively to the search for its own identity, taking with it all or part of philosophy and several sciences (curiously referred to as the humanities) and charting the space of what we now refer to, using a word of which the Romantics were particularly fond, as 'theory'" (1988, 16). But in the Colombian case, crisis names not a temporary moment to be averted but rather the permanent postponement of the possibility of establishing a sense of political and expressive foundation for the nation that is replaced by the permanent presence of the reality or threat of civil war. This appears not only in Vergara y Vergara's work but in others' work as well.

As stated previously, Jorge Isaacs and Candelario Obeso were soldiers fighting patriotic wars in the name of the radical liberals, the opposite side of Vergara y Vergara's political affiliation. Toward the end of Jorge Isaacs's life, he wrote, in a letter to a friend, of the easy slippage between the world of war and the world of letters:

On the 31st of August [1876] I fought as captain of the [battalion] "Za-
padores" in the battle of Los Chancos. When I forced the pass of Otún,
on November 13, 1876, with two battalions of the third division and "the
fourteenth of María," in order to upset the defenders of the riverbanks of
the Otún, . . . I was major sergeant and chief of operations (*jefe del estado
mayor*) of the third division of the armies in the south . . . I finished the
campaign with the recuperation of Popayán the 26th of April of 1877. I
returned with fervor to the labors of public instruction without changing my soldier's shirt, which was the only wealth that I ever got from the
campaign. (Isaacs cited by Pérez Silva 1996, 18)

The presence of "artist-soldiers," a term originally used by lawyer and scholar
José María Samper in one of his theater pieces in the nineteenth century, was
not a rarity (Ceballos Gómez 2005). In several Latin American countries in
this period, "political citizenship was closely associated with participation in
the militia" (Sabato 2001, 1312). But in Colombia war was not simply the background onto which these scholar-soldiers inscribed their literary ideals. It was
an intrinsic and constant part of their lives. By the early twentieth century, if
not by this period, as stated by Marco Palacios, "collective fratricide" had become a source of nationalism (1995). If the professional practices of these men
were heterogeneous and the practices of inscription with which these writers
penned their listenings were highly different from each other, the idea of war
as a permanent threat or as a real lived experience was a constant. The movement between war and other facets of life did not even involve, as Isaacs states
it, a simple change of shirt. War here then does not name a state of crisis, even
though war's events might lead to a total disarray of one's personal life and of
public life. Rather, it names what became the taken-for-granted and therefore
a condition of the theoretical, the political, and the literary.

Following the Benjaminian injunction that "the tradition of the oppressed
teaches us that the 'state of emergency' in which we live is not the exception
but the rule" and that "we must attain to a conception of history that is in keeping with this insight" (Benjamin [1955] 1968, 257) then we perhaps can rethink
Lacoue-Labarthe's foundational adscription of the invention of "theory" to
the emergence of the literary in the Romantic period. Here, in an impossible
to pacify New Granada, the condition of emergence of theory is rather the
impossibility of separating the state of exception from the literary. If war or the
threat of war creates a foundational political void in which the nation appears
as a permanent postponement, then this canonization of the origins of "poetry" as a failed effort to locate an appropriate (epic) song also casts Colombia

as a nation whose proper expression is always yet to make its appearance, an impossibly (violent) futurity of the past. The concept of theory that unfolds here implies an exploration of the local that, unlike Europe in the nineteenth century, rests on *not* separating the philological from the search for a local popular song and from the production of literary theory.

As we shall explore in the rest of this chapter, such a theoretical lack of distinction rests, partially, on the transposition, through orthography, of a "creative misunderstanding" of medieval and Renaissance poetics through a purposeful conceptual mishearing in the nineteenth century: the difficulty of clearly distinguishing between *cantar* (to sing) and *decir* (to say) as separate practices of enunciation. If in the Middle Ages the complexity of modes of enunciation gave rise to questions regarding theories of rhetoric (Gómez-Bravo 1999) in New Granada, the theoretical lack of distinction between *decir* and *cantar* was absorbed into literary theory by the crucial place given to song in articulating the significance of the literary and by the technology of orthography as acoustic inscription. Such a resignification of the feedback between the two terms was also partially based on the fact that practices of reading were not necessarily silent in New Granada in the nineteenth century (Silva 2009). Written texts were often meant to be heard, and the reader often understood by listening, either to him- or herself or to others read out loud.[12] A reader then can also be someone who *listens* to a text that is read out loud. "Theory" here names a technology of inscription, the practices of enunciation and listening to such texts, a particular mode of "sampling" the local, distributed between literary and musical history, as well as the modes of conceptualizing the interrelationship between them.

We can then rethink the theoretical role of Vergara y Vergara's repeated complaints of what New Granada lacked, as a foundational "deficit of originals and originality" (Richard 1993, 212) that turns the production of theory into a practice of appropriation, derivations and substitution. If Colombia is found lacking in originary texts, it is not found lacking in philological theory. The gap between audile techniques and lettered inscription as that which leads to "writing badly a knowledge that is simultaneously exalted" (Ramos [1989] 2003, 39) becomes the central locus of conceptual articulation of a foundational poetics. As such, the phoné names not only the potential ambivalence between signifier and signified but the heightened acoustic excess of the impossibility of fully separating *cantar* and *decir*, the author/transcriber and the singer/speaker, or the creative potential of mishearing from a foundational literary/musicological theory. I now turn to different ways in which such politics of foundational inscription were enacted in the nineteenth century by Vergara y Vergara.

The last chapter of Vergara y Vergara's book is titled *Poesía popular, carácter nacional, conclusión* (Popular poetry, national character, conclusion). Unlike the other sections of the book, this one deals with Vergara y Vergara's present. This turn to the contemporaneous implies addressing questions of relationality, belonging, and personhood in the nation as the entry point into the formulation of the "popular." He opens the final chapter of his book by affirming, predictably by now, that the popular poetry of New Granada is "very poor" in comparison to that of Spain but "considered in abstract still presents some interesting elements and demonstrates some intellectual richness in the 'low peoples' (*bajo pueblo*) of New Granada" (Vergara y Vergara 1867, 508). The historiographical problem Vergara y Vergara faces is that of turning a colonial legacy of "popular poetry" into a national one. He does this by defining song (or popular poetry) as a phenomenon identifiable as national due to the linguistic unity provided by the Spanish language. But he still needs to deal with the diversity of peoples within the nation. He does so by classifying song types according to two main typological principles: on the one hand, race (*raza*) and the relation between peoples (*pueblos*) and on the other, the political division of states (*estados*) within the nation. In the midst of doing so, he manages to provide a model for something that folklorists by the early twentieth century in Colombia had learned to do very well: turn the study of an alluring expressive diversity into a stereotypical description of traits that presupposes a straightforward relation between "identity," "place," and the characteristics of a particular "music" (or poetry in the language of the period). This transforms creativity and its allure into a model for mechanical reproduction *as* aura, a powerful hierarchy meant to replicate expectations of what people *should* be and do and how expressive forms *should* sound, codified as a racialization of "tradition." This presupposes that among certain peoples in the world, culture as convention is a given, something people *have* and therefore can be studied through specific objects (Wagner [1975] 1981). Their conventionalized expressive culture then functions as either a resource for identity and identity as the primary parameter of definition of the person. It is a story about the production of "normal personhoods" (Asad 1993) as part of the project of the governmentalization of diversity.

The counterinvention of the classical or traditional in the midst of the constitution of the modern has become an identifiable, repeated trait of the postcolony (Weidman 2006). Despite decades of deconstruction of the politics of traditionalization, scholars often express frustration with the eternal return of the assumptions of purity and clear origins that surround the politics of heritage (Bauman and Briggs 2003; Sandroni 2010). Such practices of purification as heritage often exceed their official institutionalization and simultaneously inhabit the realm of desire, affect, and counterproduction of political spaces such as those mobilized by social movements, the narratives of historicization that surround certain traditions (Weidman 2006), and different ritual and religious practices (Matory 2005). So the cultural politics of contemporary peoples historically assigned the role of tradition in the history of the colonial inhabit the contradictory political obligations of the place assigned to tradition in the public sphere generated by modern colonial history, peoples' own understandings of their practices, and the search for political alternatives embodied by social movements (Povinelli 2002). The idea of tradition associated with orality, then, is located in the complex politics between the reification of particular (mis)understandings of the idea of culture (and authenticity and tradition) and the reclaiming of such traditions to name other forms of knowledge-making and constitution of the self. The political injunction to tradition is generated then between multiple understandings of difference produced by the demand of rights of non-Westerners within the Western public sphere through what is generally understood in today's world as a "politics of identity" (or its aftermath). Those demands have to do, in part, with the way race and certain notions of culture were used in the nineteenth century as terms to erase, silence, or hide different ontological understandings of personhood in the expansion of Christianity as a secular, rational, modern project (Asad 1993, 2003; Anidjar 2002, 2008). The competing obligations assigned to the notion of tradition today are related to such a history. I begin by exploring how race is parlayed into culture.

POPULAR SONG AS A RACIALIZED CATEGORY OF BEING

As we saw in the previous section, for Vergara y Vergara the emergence of Spanish as the conquering language over indigenous tongues was central for positing an oral and aural unity of the nation. With that accomplished, "at the beginning of the eighteenth century all our people (pueblo) spoke a Castilian as pure as that of the people of Castile" (1867, 508). But such pure Castilian Spanish was not enough for popular song to emerge:

> Without common traditions, poetry could not become popular; neither the indigenous race nor the white one could be sympathetic to the songs of blacks; nor the blacks to the Spanish traditions of their masters or the vague remembrances of the Indians. These three races

confounded in the same territory could not see it as their fatherland, because the blacks and whites thought respectively of theirs; and the moral fatherland of the Indians had disappeared amongst mountains of cadavers; the physical fatherland, the ground on which they walked, was as foreign to them as for the blacks, their peers (*compañeros*) in slavery and misery. On the other hand, and despite the misfortune they share, the Indians and blacks reject each other in their characters and inclinations. The black intoned, silently (*por lo bajo*), songs (*cantares*) that the Indian did not repeat and vice versa; the white would sing his *romances* and *coplas*, which the black and the Indian would repeat badly (*que repetían a medias*) and only when they found analogous situations to their states of mind (*ánimos*) or to the intelligible expression of feelings and passions that are common to all men. Since this heterogeneous people did not have a previous history, proper to the country in which they all came together, then poetry could not become popular. (Vergara y Vergara 1867, 509)

For Vergara y Vergara the primary idea on which the notions of tradition, the popular, and the fatherland are built is that of race, which is understood as a shared heritage determined by a common origin. After the above passage Vergara y Vergara states that when "finally, remembrances were united" with the passage of time, another problem remained: that of "the antipathy between the races" (510). For this "obstacle to be overcome," a great deal of time was needed. Nevertheless, by the moment Vergara y Vergara was writing, "the dominated races have celebrated a tacit transaction with the dominant one" because the former have taken from the latter "the really simple and very popular songs, that is, the spontaneous ones, that describe the agitations of feeling (*las agitaciones del ánimo*), sadness, jealousy, joyous love" (510–11). Common passions and a common language solve the problem of unity posed by "the antipathy of the races." If epic song was meant to provide the narration of history, "spontaneous" songs of common feelings were meant to provide the social contract. If heartfelt expression was what provided unity in the diversity of races, then Vergara y Vergara needed a genre that somehow represented the idealized mestizaje of a proper mixture of races, heartfelt expression, and the Spanish language. Despite its African ascendancy (according to Vergara y Vergara identified as such by Jorge Isaacs as originating in the "African tribe of Bambouk") the bambuco "is the only one of our things that truly contains the *soul* and *air* of the fatherland" (emphasis mine) (512). If heterogeneous origins and heritages divide, then common affect, expressed through a common air, provided the much needed constancy of the (national) soul. An idealized form of the bambuco came to be identified as Colombia's national genre by folklorists throughout the late nineteenth and early twentieth century, as emblematic of an idealized mestizaje that emerged from a patriotic racialized historiography, as posited initially, according to Miñana Blasco (1997), by Vergara y Vergara.[13] But what I find more significant than the bambuco's never fully accomplished whitening through the ideal of a proper musical mestizaje,[14] is the need for creating a unified genre (*air*) that could express the unity of the (national) *soul*. A constancy of the soul (Viveiros de Castro 2011) needs a constancy of genre.

As stated by Viveiros de Castro, "the differentiating mechanism in European ethnocentrism is cultures whereas what is given is 'nature'. . . . European praxis consists in 'making souls' (and differentiating cultures) from a given corporeal-material background (nature)" (2010, 29–30).[15] And he adds in a footnote: "the old soul has received new names and now moves forward under a mask: culture, the symbolic, the spirit in the sense invoked by *mind*" (29). Such "making of souls" as a differentiating dimension is what is invoked here in making *genre* that which materially embodies such a soul. The type of epistemological labor that Vergara y Vergara does in the above passages is to postulate "nature" as reality and "culture" as artifice. As explained by Wagner:

> It is generally assumed that our Culture, with its science and its technology, operates by measuring, predicting and harnessing a world of natural "forces." But in fact the whole measure of conventional controls, our "knowledge," our literatures of scientific and artistic achievement, our arsenal of productive technique, is a set of devices for the *invention* of a natural and phenomenal world . . . The significant aspect of this invention, its *conventional* aspect, is that its product has to be taken *very seriously*, so that it is not invention at all but *reality*. If the inventor keeps this seriousness firmly in mind (as a "safety rule," if for no other reason) while doing his job of measuring, predicting or harnessing, then the resulting experience of "nature" will sustain his conventional distinctions. ([1975] 1981, 71–72)

This is, then, the epistemological mechanism of constructing a multiculturalism based on objectifying musical genres, person-made products that stand for the differentiating mechanism of the soul of a people against which to counterproduce the givenness of "nature." Vergara y Vergara does this through a process of assigning dispositions and characteristics to each "race" while simultaneously mutating how "race" is defined by making the dispositions of each "race" stand for its "soul" expressed in its "airs." Thus, beyond the idealized musical and racial mestizaje that incarnates the soul of the nation in the bambuco, what we find are "types" that primarily reflect particular ascendancies, particular modes of poetry and singing that are peculiar to either the white or black "race" (since the Indians, according to him, no longer sing) or to peoples from a region. Vergara y Vergara spends the rest of the chapter, on the one hand, identifying the ways whites and blacks sing, and, on the other, pairing particular regions and particular "types" with specific genres and characteristics of song or, in many cases, simply with typical dispositions evidenced in particular "character types." The project is clear: each region has a particular race or mixture of races that is expressed in particular character types and therefore yields different modes of singing or of cultural expressivity.

First, song heritage is divided into "three parts: Spanish coplas of pure origin, adopted and popularized, sung by all the races, who believe them to be their own; Spanish coplas and *romances* combined, sung by the peoples from the Eastern plains (*llaneros*), which is a population that is quite pure in its blood; and African coplas that have become popularized with their dances, and that have been adopted by the Spanish race and with even more reason, by the mestizo race (Vergara y Vergara 1867, 518). Among the people from different regions of Colombia, "the Llaneros (peoples from the Eastern Plains) are the only people amongst us that have their special poetry . . . it is the same popular *romance* as in Spain and it always contains the narration of some heroic feat" (518). Vergara y Vergara transcribes, "for the first time," the lyrics to a long *galerón* from the Eastern Plains, citing it as an example of the local forms of Spanish *romance*, and later proceeds to trace coplas that can be found simultaneously in Emilio Lafuente y Alcántara's collection of nineteenth-century Spanish popular song as well as in New Granada.[16] He thus follows or initiates, for Colombia, a disciplinary practice that became central to early twentieth-century Latin American folkloristics, of tracing local popular song verses back to similar or equal samples to be found in the Spanish *coplerío* (collections of popular verse forms). The existence of a proper popular song is expressed mainly by a successful type of *linguistic* mimesis of the Spanish copla and *romance*. The only local forms that accomplish this are the popular songs from the Eastern Plains, sung by "a population that is quite pure in its blood." Here, racial affiliation is mentioned not in terms of common origin understood as "coming from the same place" as earlier but in terms of blood purity, a corporeal "natural" substance. If constancy of the soul is expressed in the constancy of genre, "race" is a problematic concept because it lies between the body (as blood) and the soul, constituted as much by ideas about nature as about culture (Wade 2002), an issue to which we will return.

On the other hand, "black poetry" is characterized by a failed or badly done linguistic mimesis but also by an outstanding musicality, which Vergara y Vergara identifies as the "gift" blacks bring to whites:

> It is a rather strange thing that cultured whites (*cultos blancos*) have not been able to give even one single happiness (*alegría*) to the blacks; and that these, banished from their lands and foreign, have brought such a gift to the whites. They have only received from the whites, some *coplas*, and that because they have had to forget their African dialects. The black race, acclimatized in its exile, is eminently poetic, and above all, a musical and singing one (*música y cantora*); their voices are marvelous in their elasticity, expansion and harmony . . . The blacks sing our Castilian verses in their *bundes* and their bambucos, and conserve some peculiar songs that they sing in their very beautiful voice with airs that they remember or invent, successfully overcoming the most enormous difficulties of song and music. There is no need to write down some examples of coplas adopted by them, because any Spanish reader knows them, with rare exceptions. We will rather give some examples of the poetry that is characteristic of black peoples. (Vergara y Vergara 1867, 513–14)

The examples of black poetry that Vergara y Vergara gives are characterized by, on the one hand, a responsorial structure that he indicates on the page (alternating duets and solos) and, on the other, a mispronunciation that appears as a misspelling that yields a nonstandard Spanish, characterized mainly by incomplete words (*ñor* instead of *señor*), the elision of consonants at the end of words (*vendé* instead of *vender*) or between syllables (*sumecé* instead of *sumercé*). For Vergara y Vergara such mispronunciation is the sign of a lack of ear for different languages manifested as well in blacks' penchant to turn foreign languages into onomatopoetic nonsense (513).

For him, then, black peoples have a gift for music but the language they use is badly pronounced Spanish due to the loss of African "dialects." Blacks have a voice, but they do not know how to *say* well (*bien decir*) and potentially, their talk is nonsense due to mistranslation. In a country that by the end of the nineteenth century would be obsessed with the implementation of an educational politics of proper pronunciation as a sign of nationalist distinction and good governmentality (see chapter 4), such ideas amount to an exclusion from political participation. The failed phoné of blacks is a diglossia that acoustically delimits their linguistic competence while recognizing the musicality of the voice. It is as if blacks were capable of a profound authenticity (in the gift of "happiness" and acoustic vocality that their music brings) but not of fidelity (in the double sense of perfect acoustic imitation and loyalty/servitude) because they are incapable of reproducing language well.

So voice works on a double register: on the one hand, voice references an acousticity that is indexical of musical worth, and on the other, it references a failed linguistic mimesis through an ideology of language articulated in writing through misspelling (Ramos [1989] 2003). Such misspelling intends to construe mispronunciation as the "iconicity that allows a person to identify the discourse as a voice by virtue of that voice being recognizable as a generalizable figure of a certain type" (Keane 2011, 174). Vergara y Vergara turns black vocality into a moral figure, a type. As stated by Keane: "Once a certain way of speaking begins to circulate across contexts and unite (to be recognizable across) different moments of interaction in such way as to produce the effect of a person having a certain knowable character, of a certain moral inflection and socially identifiable nature, then it takes on the full-fledged nature of a social figure or stereotype" (Keane 2011, 174). So Indians are silenced/t, blacks musically virtuosic yet linguistically inept, whites linguistically superior. "The moral figure, as a type," stands for the soul, and the description of traits becomes the definition of personhood — the differential making of souls.

But this is just the first step. Bad pronunciation is a feature that characterizes other people as well. In a country of "vulgar" mestizajes who were yet to be purified themselves through whitening, this type of orthographic inscription of bad pronunciation was also extended to other commoners in the type of writing called *costumbrismo* that prevailed in nineteenth-century Ibero-America. This required a move from race to region, which Vergara y Vergara proceeds to do at the end of the chapter. He closes the book by describing "the character that is proper to (*propio*) the peoples that today form the set (*conjunto*) of what today is the republic of Colombia" (Vergara y Vergara 1867, 524). In each of the states of Colombia the idea of regional races or types is described through specific characteristics and customs. Thus he speaks, for example, of an "original type, the goajiro . . . who does not figure among the civilized races but is no longer part of the barbaric ones" (525) or "the pastuso cultivates agriculture and arts: is a craftsman and a painter but is not a poet or an orator or a writer" (527); or yet another example: "the people from Valle del Cauca are divided between those [originally] from Popayán found among the high class of Cali and Buga, but modified by the warmer climate and by life in the fields; in the mestizo type, product of the black and white race that today is the ferment of the state in its political concerns, its great pretensions and its scarce love of work; but when it combines more and becomes more extended it will be a great people; and the black morally abject due to slavery, and arrogant and spiteful due to the first days of liberty" (527). And so on for each state. Thus we have a movement from race as common origins, at the beginning of the final chapter of Vergara's book, to race as masked in the notion of character type of each geographical region, in which character type embodies the soul of a people. The new ingredient introduced in this final twist is that of the relationship between geography and "character types" as the central site of a racialized expressivity, a crucial link to botanist Francisco José de Caldas (Von der Walde 2007).

Throughout the eighteenth century, the Spanish sought to curtail the power of Creoles in Spanish America through a series of measures known as the Bourbon reforms, a policy that ultimately alienated Creoles and was crucial in motivating the struggle for independence. A central element of such reforms was financing botanical expeditions in a global race between competing European colonial powers to control natural products. The botanical expedition of the New Kingdom of Granada (1783–1816), was led by world-renowned naturalist Jose Celestino Mutis, a Spaniard who settled for life in New Granada, and whose most outstanding Creole disciple was Francisco José de Caldas (1768–1816).[17] The historical reappraisal of Caldas's work in recent years has placed him at the center of a controversy that credits him as the author of the idea of the interdependency between organisms, altitude, and climate in the torrid zone, a notion historically assigned to Humboldt and central to his writings on the "geography of plants."[18] The idea of climate as one of the major influences on living beings and therefore on human "character" was hotly debated in the journal *El Semanario del Nuevo Reyno de Granada,* founded and published by Caldas in Bogotá between 1808 and 1811 (Nieto Olarte 2007). A large part of the debate fluctuated around whether it was primarily climate or education that formed the virtues and vices of men (Nieto Olarte et al. 2005). For Vergara y Vergara it is obviously climate, as evidenced by the above passage.

At the beginning of the nineteenth century, the illustrated Creoles of New Granada understood "geography" as a "wide field of knowledge that included issues regarding the economy, population, climate and natural resources" at the service of governmentality (Nieto Olarte 2007, 127). For Caldas, one of the purposes of economic geography was to learn about "the genius, the customs of its inhabitants, their spontaneous productions and those for which it finds a home in the arts (*y las que puede domiciliar en el arte*)" (Caldas cited by Nieto Olarte 2007, 127). Caldas saw the geographical variety of Colombia as ideal for this type of study for "there are few places in the globe with so many advantages to observe, and we could even say, that touch upon the influx of climate and food on the physical constitution of man, his character, his virtue and vices" (cited by Nieto et al. 2005, 98). But he also argues in his later writings (those of 1810 in the midst of the battles of independence), the "degeneration" of indigenous peoples and Africans was due to the bad government of Spain.[19] According to Erna von der Walde, Vergara y Vegara's literary history was meant to "create a genealogy and archaeology of Francisco José de Caldas as a transitional figure" that permitted a connection between a Hispanic colonial past and a Hispanist project of the nation (2007, 250). Such a geography of character types made its full appearance not only in song but also in a literary genre called the *cuadro de costumbres* (custom sketches).

Vergara y Vergara's depiction of black music described above is inserted into picturesque depictions of the way slaves sang in sugar mills and in church in the style of the cuadro de costumbres, one of the main genres cultivated by Vergara y Vergara and the figures associated with El Mosaico. According to José Manuel Marroquín, one of the members of this group, an objective of the genre was "to depict, for the instruction of strangers and for posterity, the customs of countries in specific moments. It can also be composed with the objective of correcting reprehensible or defective elements in such customs (Marroquín cited by Duffey, 1956, xii).[20] Costumbrismo writing involved a wide array of literary practices "from variants of the social picaresque novel adopted from the Spanish model of Larra and Mesonero Romanos, to the musings of social and political practices, to travel chronicles written in a descriptive format that was halfway between the scientific and the narrative" (Von der Walde 2007, 248). If "custom" or "character" is the name given to "the social form of natural impulses" (Sahlins 2008, 41), the proper organization of such "custom" through a national literature would transform undesirable mores into picturesque virtues. The particular aesthetic of costumbrismo as representative of a region and a people is accomplished through a "cultural rigidity assigned to the local" (Rama [1984] 2007, 32). It is less a "phonography of popular speech" (49) than a "reconstruction suggested by the management of a regional lexicon, dialectical phonetic deformations, and, to a lesser extent, local syntactic constructions" (48). Ultimately then, the idea of race is located between the differentiating soul and elements of nature: blood, land, and climate. Thus song-making and the literary description of customs lies between the "scientific" and the "literary," between the given (nature) and that which differentiates (culture).

For Peter Wade, "ideas about persons, identity and belonging traffic back and forth between the apparently separate domains called nature and culture, unsettling their boundaries, overlapping their radii of action" (2007, 11). Vergara y Vergara's racializing discourse moves from kinship as common origins, to kinship as demonstration of blood purity (prevalent in eighteenth-century Colombia), to defining the "mutuality of being" (Sahlins 2013) between persons through the relationship between territory and language. According to Porqueres i Gené, writing about Basque nationalism and race, the move from the manipulation of notions of the person as primarily defined by a "right of blood" to one primarily defined by a "right of soil" is politically motivated by "the crisis of the genealogical model of the nation, engendered by the very logic inscribed in the rhetoric that constitutes it, which opens the way to new identitarian formulations in linguistic-territorial keys" (2002, 51). Thus, "blood and soil far from being opposed, appear as variants of a single theme, that of the presence of the ancestors in the definition of the nation. Land thus becomes a mediator between those who existed in the past and those who currently live on it" (Porqueres i Gené 2002, 60, cited and translated by Wade 2007, 12). When blood purity could no longer be used to account for socio-racial differentiations and hierarchies, such as happened in Colombia with the intensification of mestizaje in the eighteenth century and with the move from a colonial language of castes to one of organization of diversity in the unity of the nation, language and territory became key elements to do so. Such a change was also accompanied by a transformation in the epistemological work demanded of sensory perception (Smith 2006), a figure invoked by Vergara y Vergara through the turn to "common feelings" evoked by song. When the "natural antipathy" between races is replaced by a common passion evoked by song, "no longer is passion fighting passion. The nation is the passion—the body politics of the body politic" (Sahlins 2008, 82).

Even if framed scientifically and in literary terms, this is also a concrete theory of affect. A "mobilization of modern desire" (Palmié 2002, 482) reconfigured as the "gift of music and dance" recasts the negative physiologies of blacks and mulattoes not as disgust, but as an allure.[21] The upholding of "boundaries of race and reason" through a scientific and literary discourse of affect helped create the distinction between a bourgeois cultivated autonomous self and the "irresistible" physiology of a racialized other (Palmié 2002, 278).

This epistemological move was also crucial in creating the difference between an autonomous Western classical music and a "popular" or "traditional" one marked by the allure of its exotic acoustic physiology, a "carnal musicology" (Le Guin 2006) mapped onto the production of othered bodies.[22] The emergence of an idea of Western music as an autonomous art during the early nineteenth century was possible because other musics and aural spheres (notably language/voice as "orality" and music as present in the body of others through voice and dance) could carry the embodied theories of affect and music as language that had characterized understandings of Western art music until the eighteenth century in Europe.

The Romantic period was characterized by the realignment between theories of music as a language of affect and expressivity, on the one hand, and self-contained theories of music as an autonomous transcendental (and mathematical) art, on the other (Chua 1999; Neubauer 1986). This transformation coalesced through the emergence of the idea of the musical work around 1800 (Goehr 1992). According to Goehr, prior to the mid to late eighteenth century, theorists constructed musical meaning "from 'outside' of itself, deriving as it did from music's ability to influence and empower a person's religious and moral beliefs, or from its ability to imitate the nature of persons and the world" (152–53). This "external significance" of music was further accomplished through words because, unlike melodies, they were "intelligible," and had "concrete and specific semantic content" that could be harnessed in the service of representation casting purely instrumental music as practically worthless (153). But under the new aesthetic, which Goehr locates as emerging around 1800, "the significance of fine art lies not in its service to particularized goals of a moral or religious sort, or in its ability to inspire particular feelings or to imitate worldly phenomena. It lies, rather, in its ability to probe and reveal the higher world of universal, eternal truth" (153). Thus, the consolidation of the notion of the "musical work" emerged through a "*transcendent* move from the worldly and particular to the spiritual and universal" and a "*formalist* move which brought meaning from music's outside into its inside" (153). Since the idea of "fine arts" was tied to the ineffable, instrumental music, a medium with no "particularized content" (153) became the ideal art to manifest this.

Neubauer associates the emergence of this "revolution" "induced by the emergence of classical instrumental music" (1986, 2) to a "Romantic revival of Pythagoreanism" (7) and mathematical "non-representational" approaches to music in general. He contrasts two longue durée traditions in the understanding of music in the West. On the one hand, he mentions a mathematical one, "from Pythagorean notions of harmony in antiquity, through theories of music as a mathematical science in the Middle Ages, Renaissance polyphony, seventeenth-century Pythagorean theories of a musical cosmos, rationalist theories of music in the Enlightenment, to the Romantic revival of Pythagoreanism . . . down to computer and serial music" (7). On the other hand, and equally important, were the "verbal or rhetorical approaches to music, of which affect or expressive theory are but variants" (7). For him, the "alternating dominance, the frequent battles, and the occasional peaceful coexistence of verbal and mathematical approaches to music constitute the history of music theory" (7–8). So the association of music with transcendental notions of the ineffable implies a simultaneous mathematicalization of the work of art, locating its analytical domain in a nonrepresentational sphere. This is seen as part of the division of the world into a modern disenchanted one and an enchanted one, a process that Chua traces as occurring through several musical instances from the sixteenth century onward, but gaining its full manifestation in the nineteenth century through "the expulsion of music from language, as if tones and words were separate entities vying for power" (Chua 1999, 23). This also involves the interiorization of the subject understood as an individual upon whom fell the rational discernment of the sense of such (dis)enchantment (Moreno 2004). In such Romantic discourse, as instrumental music was being hailed as the ultimate autonomous art, vocal music and voice came to stand for the soul of the self (Chua 1999).

This reorganization of the idea of voice as a sign of the soul of the self and instrumental music as the most autonomous of arts is the sign of modern music and of musicology par excellence. It is no coincidence that the epistemological emergence of orality, as well as that of embodied musical others, arises at the same historical moment as the idea of autonomy in Western art music. Different musical ontologies and epistemologies were split between the (irrational) othered quasi-objects and quasi-bodies of Afro-derived musicalities, silenced/t Amerindians (or indigeneities), and transcendent Western art musics as different types of embodiment of the soul or the spirit. Here musical works and performers historically conceived as quasi-objects under slavery (in the Latour sense of the word) (Palmié 2002) are redeployed to the (ethno) musicological playing fields of world music. Thus, the division between music and world music (as orality) is complementary to the emergence of the idea of the musical work that culminates in the nineteenth century. While the idea of absolute music came to stand for the notion of music itself (Chua 1999), "orality" was attuned, via a politics of differentiation of the soul, to the affective, bodily, and rhetorical theories of expressivity as (musical) language. The simultaneous rise of the idea of the autonomous musical work and the folkloric object neatly divides the understanding of regimes of art and musical products, in the same period in which, as stated by Palmié (2002), the relations between race in the colonies and class difference as a rural/urban divide in the metropolises was being rearticulated. A radical redistribution of the senses that separates "popular song" from musical (and musicological) aesthetics, a separation made complete in the early twentieth century when song became the ideal genre/product to be inscribed through the gramophone, a technology that could only handle small musical forms. Thus, the separation of popular song and music is not an effect of the rise of musical technology; it precedes it in the separation between song as popular (product) and music as (work of) art, preparing the way for the rise of song as the main carrier of mass music.

Yet let us return to the politics of the voice. If, as stated at the beginning of this chapter, all hearing is ultimately mishearing, bringing nonsense into sense, then the misspelling that bears the mark of such a process could also be used to signal a different politics from that enacted by Vergara y Vergara and costumbrismo. The same orthographic technique used for the inscription of acoustic stereotypes in costumbrismo writing could be used for voicing different understandings of the relation between self, expressivity, song, and voice. I turn now to the work of writers Candelario Obeso and Jorge Isaacs who, while valued by the intelligentsia in Bogotá, differed from the Catholic conservative project of El Mosaico and of Vergara y Vergara's intellectual circle in crucial ways.

Candelario Obeso's Phonography

In 1877 Candelario Obeso (1849–1884) published an unusual book called *Cantos populares de mi tierra* (Popular Songs of My Land). The book consists of sixteen original poems meant to transcribe, through the use of nonstandard orthography, the accent and modes of speaking of the peoples from the Magdalena River basin in the Colombian Caribbean where Obeso was born. The combination of original poetry, an acute hearing inscribed in a detailed, almost illegible alternative spelling and the careful instructions given at the beginning of the book for reading the poems out loud, highlight his attunement to the sound of language. His work anticipates, by several decades, the exaltation of the sonorous dimensions of Afro-Caribbean pronunciation that would be one of the poetic signs of the Négritude movement (Maglia 2010). Graciela Maglia sees Obeso as a writer who is ahead of his times in two senses: by proposing Caribbean popular culture as national in the midst of a scholarly project that sought to reduce its significance and by aesthetically "confronting the literary canon of the times" through the formal particularities of *Cantos populares de mi tierra* (Maglia 2010, 10).

Born in Mompox, Obeso has been hailed as the poet of the bogas, and the first Colombian Afrodescendant writer (Prescott 1985). He also wrote grammars and theater pieces, translated texts of applied linguistics for the teaching of Italian, English, and French in Colombia as well as military texts, and wrote collections of poetry beyond those that comprise his famous songbook.[23] As a black grammarian from the Colombian Caribbean who spent most of his adult life amidst the Creole lettered elite of Andean Bogotá during the years of consolidation of the Catholic Hispanist national project, Obeso stands out as an exceptional figure, one whose singularity defies comfortable niches of interpretation of his work.

Obeso himself and the significance of his work have been analyzed mostly from the perspective of his complex emplacement in the lettered city. His work and persona have been alternately seen as either resisting such a project (Maglia 2010) or as resisting it but tragically trapped within it (Jáuregui 2007). In Graciela Maglia's view, Obeso appears as exemplary of both an artistic vanguard and an unusual ethnographic sensibility. He is a "social poet that brings art to life, inscribes difference and creates a new canon from popular culture, with a new notion of beauty constructed from the bases" (Maglia 2010, 20). She sees his use of misspelling as part of the "linguistic variation characteristic of Creole American tongues" (21). Through such language, Obeso celebrates the culture of runaway slaves (*cimarronaje*) and a "life retired into the backlands, a libertarian topos par excellence" (22). For her, Obeso "writes from an autonomous position in relation to the academy and the political world of the times, and they answer with the rejection of omission" (25). Other authors see him as trapped between the lettered city and the world from which he came. The resistant but tragic interpretation of his work emphasizes that "Obeso is trapped between the black culture to which he belongs and the 'white' writings from which he expresses himself" (Martín-Barbero cited by Jáuregui 2007, 53). And historian Jorge Orlando Melo brings out yet another facet of Obeso's strange emplacement in the lettered city. To him Obeso's *Cantos populares de mi tierra,* "suggests the equivocal idea that it could be a compilation he had not written. Behind such highly polished works . . . was the idea of imitating what he had heard all his life: the popular tongue, the popular accent of a very difficult phonetics . . . but writing sixteen poems with the intention of imitating popular phonetics was something totally exceptional, even as a scientific attitude and not only as a poetic attitude" (Melo 2005, 12–13).

Cantos is then an equivocal piece and Obeso an uncanny figure, one that defies a comfortable adscription in the politics of recognition. Neither fully belonging to the lettered city nor to the realms of the proper documentation of the popular, Obeso creates a poetics that seeks to imbue the lettered word with sound. He engages in a politics of inscription that highlights the arbitrary nature of standard orthography, and emphasizes alternative pronunciation as the site of another knowledge. His phonography is one of transcription rather than inscription in that it seeks to carefully document through a highly acousticized orthography, the sounds he heard, knew, and pronounced onto the page. He even gives his readers instructions on how to pronounce the systematic alteration of spelling he chooses.[24]

His is a purposeful poetics of alternative transcription of sound in the lettered word that hears beauty and "elegance" of diction there where the lettered elite only saw mispronunciation as custom. Says Obeso, "in popular poetry there is and always has been, without the philological advantages, a great abundance of delicate sentiment," one which he describes as having "metaphorical intelligence" and as "essentially poetic" (Obeso edited by Maglia 2010, 66–67). His proposal appears as one that challenges the mimetic obsession of the colony: "only with careful cultivation do nations come to set the foundation of their positive and true literature . . . may the youth who love progress work with this purpose, and that way, the sad furor of imitation, that has been so detrimental to Hispanic American letters, will soon calm down" (67).

But the critique of power in his work lies not only in exalting beauty and intelligence where others only see custom but also in the way he highlights mishearing, making it difficult for others to follow his instructions for reading the poem aloud, through a transcription that is challenging for any native Spanish speaker to read. On September 5, 1877, Colombian philologist Ezequiel Uricoechea wrote a letter from Paris to his friend the Colombian philologist Rufino José Cuervo, who was then in Bogotá: "I have read with much interest the costeño verses and I thank you for your gift. By following the method of reading that you indicated to me, I have only missed about three words" (Uricoechea 1976, 193–94). Perhaps, but no doubt much of the acoustics were lost on his tongue. By highlighting mishearing and misunderstanding, Obeso flaunts a nonsense that the grammarians from Bogotá would be at pains to bring into sense. He writes against the refusal to listen to the popular and writes back to the lettered elite by using mishearing as a political tool. He creates a disjuncture between what is *seen* on a page, what is semantically understood, and what is impossible to pronounce back from the written page for an Andean lettered elite, making it difficult for them to bring nonsense into sense. His technique is less a misspelling and more a transcription, acoustic scores more than poetic words, a poetics of sound made evident in the difficulty of reading his poems. Obeso's world is not only that of the lettered city. It is also that of an acoustic expertise that our contemporary understanding of the term *popular* hides rather than elucidates.

As we saw, in the hands of the New Granadian literati the term *popular* comes to designate less a knowledge and more a resource. Obeso's poetics, on the other hand, designate a type of knowledge that he hears, transcribes, translates, and, even sometimes, deciphers for the lettered elite, because they are not only unable to hear or read it but also to understand it. Thus, for example, for the poem *Epropiación re no corigos* (Expropiación de unos códigos, expropriation of some codices) he provides not a *translation* of his acoustic renditions to standard Spanish but an *explanation* of the content of the poem.

The poem's narrator has been given some legal documents by a member of congress, who uses them to fix the walls of his home. Beyond naming the uselessness of the law, the poem is a powerful critique of the lettered city and its racial order. It begins by ironizing the dependence of blacks on whites through a series of comparisons of interdependence in the world of plants and trees and by feminizing that dependence through an allusion to the creationist myth of women in the Bible. It speaks about how every being has others that are more powerful and on which one has to rely to solve one's problems:

Cara sé tiene en er mundo,	Every being has in this world
Apácte re la cotilla,	Besides the rib
Otro sé que poc ma fuécte	Another that by being stronger
Ej er puntá re su vira.	Sustains his life.
Tiene er bejuco der monte	The liana in the jungle
Siempre un ácbo a que se arrima;	Has a tree on which to rest
I erte palo tiene er suelo,	And that tree has the ground
I er suelo en acgo se aficma;	And the ground affirms itself on something
Yo, branco, le tengo a uté	I, white man, have you
En uté la pena mías	In you, my troubles
Jallaron siempre consuelo	Always found consolation
I pronta la melecina	And medicine at hand

Obeso then *explains* the beginning of the poem for his lettered friends in its "castizo version": "In the world every being has, besides their beloved, another that, through the support they provide, makes existence stronger. The liana in the jungle has the tree around which it wraps itself; this tree has the globe of the Earth. And the globe of the Earth in turn holds on to something . . . I, for one, have the affection of your protection, for which I feel honored, and where my sadness always finds consolation and my suffering, relief" (Obeso in Maglia 2010, 91).

But what at the beginning of the poem appears as the gratitude of a black man being protected by a white man is revealed by the end of the poem, and through acerbic political irony, as a dismissal of the hypocritical authority of the lettered city. The poem ends with a rabid critique of the code of honor that hides racism's inequality in the supposed virtue of friendship. The lettered city appears as an inhospitable place, so much so that the only use for its piles of papers is their dissolution into the mud-walls of the narrator's own poor home. A home that absorbs its papers as substance to build walls, dissolving its lettered pretensions:

Ayer tuve en er Congreso	Yesterday I visited Congress
I me rió er dotó Ecamilla,	And Dr. Escamilla gave me
Sei volúme pa que a uté	Six volumes
Se los trujiera enseguia,	To bring immediately to you
Maj apena lo cojí	But upon grabbing them
Compré acmiron (meria libra),	I bought half a pound of starch
I vine a tapá e mi choza	And came to build the walls of my hut
Lo juraco y la j'endijas.-	To cover the holes and crevices
Si eto le parece má,	If you think this is wrong
Iré luego ar dotó Anciza;	I will go to Dr. Ancízar
er tiene er pape a montone	He has lots of paper
Si uté papé necesita;	If paper is what you need

Obeso finishes the poem with his critique of friendship as the virtue that resolves the contradictions of racism, through a series of comparisons with the natural world. Again, he provides both the poem and the explanation. I cite only the latter:

Yesterday I was at the house of chambers of which you are a member, and the doorman, Escamilla, gave me to bring to you, the Codices of the Union, that, due to my necessity I reduced to certain things that momentarily brought happiness to my very sad and miserable home. If you find this wrong, I will do all to make up for this insignificant grievance; but if it so happens, which I doubt, then I will know that friendship is not what they say it is; that the strong forest makes its nourishments scarce for the insects that it produces and that live in it; that the strongest support for a tree branch is not enough for sustaining the fruit the simple peasant entrusts to it; that the tender pigeon and the chicken, the mongrel and the pig barely differ from each other; that no one in this vale of tears and misery enjoys a real friend . . . All that I will know sir, although not for nothing. I will always be who I am and how I am, no matter how much I taste the cup of disillusion. (Obeso in Maglia 2010, 92)

Obeso disrupts the pretended distribution of knowledge in the lettered city by denouncing the hypocrisy of its intentions. This political awareness unfolds as a political irony expressed though poetic uses of his intimate knowledge of the world of plants and animals. But Obeso neither turns the exuberance of nature into paradisiacal excess, nor into the sublimation of transcendence as in the Humboldtian legacy, nor in the need to order it for the New Granadian scientific elite. While indeed, his poetry is that of resistance, rather than a symbolic outside, Obeso invokes the quotidianness of the vast territories of the nation, the spaces of cimarronaje, according to Maglia (2010), as intimacy. This intimacy becomes the material of the poem itself in its acousticity and in the poetic semblance given by him to animal and plant life. In Obeso's poetry animals are not that different from humans, not because humans have a debased "animal" nature, but because the biological cycles of life and the comportment of animals and plants metaphorically evoke the conditions of life for humans.

Such quotidianness is also present in the way his poetry is partially based on the poetics of popular song, such as his famous *Canto del boga ausente* (Song of the absent boat rower) whose refrain is also found, in a similar version, in Jorge Isaacs's novel *María*.[25] In that sense, some of the poems are more arrangements on the page than transcriptions, in the manner of composition of traditional song—a succession of eight-syllable lines used to build four-, six-, or ten-lined strophes in different rhyming schemes that alternate with repeating refrains and that can be adapted to different melodic patterns. Arrangements are interdiscursive pieces that emerge out of the conscious use of previous works or sources (Szendy 2008). They imply the transformation of originals into a new piece that may or may not be considered a new, original work. In sum, the contestatory poetics of arrangement used by Obeso that challenge the normativization of costumbrismo are based on audile techniques that the grammarian elite from Bogotá did not have, because they could not hear what he heard.

And yet, despite its contestatory nature and the way it references a knowledge that the grammarians of Bogotá could not experience, each and every single poem from *Cantos populares de mi tierra* is dedicated to the members of the lettered elite from Bogotá, as if personally calling for the attention of each one of them. The principal topic of the poems and one of the main ones of his later creative work is that of the difficulties of being black in this white order. If on the one hand, Obeso ironizes the lettered elite, on the other, he wants to be taken as an equal within their order. Jáuregui elaborates on the alienation this generates:

Cantos pretends a double translation: from orality and the "vernacular traditions" of sectors marginalized from the national process to the dominant cultural codes and to Literature; and also a translation from the "Afro" to the "national" . . . This gesture undoubtedly represents a "hybridization" of the Hispanic horizon and a questioning of the cultural univocity of nationalism of the Colombian *Republic of Grammarians*. But at the same time, he distances himself definitively from the translated cultural tradition; *Cantos* is at the margins, but not outside the *lettered city*. The Afrocommunities that are his referent produce—with other types of social practices—the social meaning of their identities outside the *lettered city* and the literary institution. In these circumstances Obeso is a tragic translator. He enters the *lettered city* translating and betraying the "songs" he exalts (*reivindica*) as well as the Hispanic grammarian tradition of national literature in which he tries to situate himself. The result is alienation from the culture he aspires to represent as well as from the *lettered city*, for whom such translation embodies an ethnolinguistic pollution or, in the best of cases, a folk curiosity. Loyal and traitor to the nation and to the cultural spaces of difference, *Cantos* is out of place everywhere. (Jáuregui 2007, 54–55)

In what sense is Obeso tragic? In the sense, says Jáuregui, of a double betrayal to the lettered city and to his own world, a seeming reference to the classical interpretation of tragedy in Oedipus in which knowledge takes the form of a crime against one's own world. But Obeso neither commits the crime nor does he refuse to acknowledge the import of his actions. Obeso is in full knowledge of his actions, to the extent that one can be, and proposes his text *as* political action, the only one that he seemingly has. Rather, it is the lettered city that betrays him by denying him his place after having given him a new technology for inscribing knowledge. The violence is perpetrated by the lettered city in its inhospitality to the auditory knowledge he brings. They are the ones who *refuse* to listen. Obeso is the postcolonial subject in full knowledge of how he is being denied entry into the new nation. This knowledge is what makes him live dangerously in the borderline between radical exclusion and desire for recognition; between refusal to work according to the lettered city's own terms (the stifling effects of costumbrismo upon his own heritage) and the full awareness of the implications of flaunting his own knowledge back to them; between irony and apparent condescension; between a celebration of himself as an outsider and the political impossibility of belonging; between the sensibility of Romanticism and the melancholia of conditions of exclusion that are impossible to transform. This is an outside that daily teeters on the brink between the dismissal of fully subscribing to the available options (exoticism, costumbrismo or enlightenment) and the realization that the desire to participate fully in the affective and intellectual order of the lettered city will never be realizable.

Obeso shows us the tragic price to be paid in the refusal to exoticize otherness as a politics of the self, to turn the exuberance of nature into the transcendence of the naturalvölker or a version of the marvelous real *avant la lettre*, or to patriotically celebrate in costumbrismo style. His is a Romantic realism that, while poetically harnessing the irony of the Colombian Caribbean, makes visible not so much the tragedy of a partial (non)belonging to the world of enlightened men, as the awry metaphysics of the impossibility of hybridity and the enormous difficulties of creolization as a project of emancipation. As if he showed us the traces left by the irresolvable cleavage of belonging to two worlds at the same time.

Obeso was an exceptional figure among the first generation of educated men who lived in the times of the abolition of slavery, even within the Colombian Caribbean where mulatto intellectuals played crucial roles in the political struggles for independence (Múnera 1998; Helg 2004; Cunin 2003). He lived in a transitional period in which the practices of a slave society persisted even though slavery had been abolished. As stated by Oscar Almario, during the early years of the postcolony, "economically, blacks, as slaves, continued to be considered as 'things,' 'properties,' or 'investments' to be preserved or recuperated by the slave owners. Politically, enslaved blacks and their descendants constituted a kind of spurious or incomplete membership in the nation" (2007, 219). Obeso's work highlights the way the price of singularity appears as an irresolvable inner schism.[26] This is what Povinelli names as the problem posed by a "radical difference," one that does not fit comfortably in any place:

> Alterity does not uniquely refer to moments of experienced or understood maximal heterogeneity across socially or culturally differentiated groups (paradigmatically found in colonial settler encounters), though we should not ignore nor shy away from the fact that fundamental differences do exist between real and imagined means and modes of producing a good life. Even when such social heterogeneity exists, what is experienced as radical difference is not interior to the social forms themselves but exterior, so to speak, or emergent in the spaces of their intersection—what George Simmel called sites of contact. Moreover, this experience may well create irresolvable cleavages not between the

two groups but within one of them, which had previously tacitly accepted and experienced itself as a collectivity. . . . Likewise the interior space of the subject may be rent, the "I" of myself lost in the field of irreconcilable moral injunctions. (Povinelli 2002, 137–38)

This radical difference is both Obeso's gift and his curse. Obeso is, along with philologist and naturalist Ezequiel Uricochea (see chapter 3) one of the best translators of modern languages in Colombia in the second half of the nineteenth century. Rather than worry about etymological purity and grammatical correctness, with Greek, Latin, and an idealized standard Spanish as the models of philological and linguistic excellence for the new nation (see chapter 4), he translated works for the contemporary world: a manual of arms, and courses for modern languages: Italian, French, English. His skill as a translator was marked not only by his capacity to move between languages and between the world of acoustic inheritance and the learned lettered one, but also his ability to enmesh such worlds in a possible future, as he so clearly asserts in the introduction to *Cantos*. In his creative use of the vanguard technology of inscription of the time (writing), he belies the understanding of blackness or Caribbean creolité as "always oppositional to technologically driven chronicles of progress" (Nelson 2002, 1). By doing so he reveals the capitalistic logic of costumbrismo (the antecedent of "folklore") as one that erases the emergent properties of artistic practices to create fixed, marketable products. In place of a dialectic of counterconvention and invention, the popular song poetics of costumbrismo propose repetition as an obsessive, anachronistic, fixed return. Such politics of time seal the past from the present (Palmié 2002). Obeso's "Afro-futurism" (Nelson 2002) defies not only this, but, as he states it, the mimetic impulse of copying European modernities in the nation's lettered elites.

Now I turn to another regional writer, Jorge Isaacs, and his modes of inscription of Afrodescendant song in the novel.

JORGE ISAACS, SONG, AND THE POETICS OF SILENCE

In the early twentieth century, folk song collection from different regions became the guiding principle of folklore research in Colombia. While songbooks would primarily be used to discipline poetics according to the dictums of the description of customs, particularly in the documentation of the copla, the inscription of acousticity into literary fiction enmeshed it into a broader narrative project. This project involved different practices of inscription based on different ideologies of the local and the consequent emergence of potential anthropological sensibilities. The dialectic between the songbook and the novel was crucial to the constitution of a history of studies of traditional and popular music in Latin America and the Caribbean that enmeshed the anthropological, the literary, and the acoustic into the reflexive stance of the novel.

In that sense, it is interesting to compare Candelario Obeso's songbook to Jorge Isaacs's popular song poetics.[27] As stated at the beginning of this chapter, Isaacs's songbook is the other known Colombian popular song collection of the nineteenth century. It was never published during his lifetime. The songbook was first edited by folklorist Guillermo Abadía Morales in 1985 and a second critical edition by María Teresa Cristina appeared in 2006. It is a collection of popular coplas written in standard Spanish, as if it were more a study of poetic form and language use than a public depiction of local lore. Thus, the collection seems to function as field notes that feed Isaacs's ethnographic sensibility rather than an exploration of costumbrismo style.

Isaacs left a relatively large body of unpublished and unfinished texts. Most of his original poetry was not published during his lifetime and two of his novels remained unfinished (Cristina 2005, 2006). After his famous novel *María* (1867) Isaacs primarily published documentary works, isolated poems and a theatrical piece. *La revolución radical en Antioquia* (1880), the book in which he finally abandons his concern with pleasing manners and adopts a frontal critical public voice, is a justification of Isaacs's political actions as a radical liberal in the state of Antioquia and a drastic political critique of the partisan struggles in that state between 1876 and 1880. This work, yet to be studied in detail, is an important historical document for understanding the schism between radical liberalism and the more conservative liberalism that emerged in this period and that eventually led to the triumph of the conservative party in the late 1880s, and, ultimately, to Isaacs's extreme political alienation toward the end of his life. *Estudio sobre las tribus indígenas del Magdalena* (published in the *Anales de instrucción publica* in 1884) is one of the earliest ethnographies to be written in Colombia (see chapter 3). Rather than an isolated gesture, then, the songbook appears as part of such a documentary impulse. Also, due to its having remained unpublished during his lifetime, it simultaneously inhabits a realm of erasure and silence that characterizes not only *María* (Avelar 2004) but, as I will argue here, the early phase of Isaacs's anthropology of popular music traditions.

María is one of the most cited books by Colombian scholars of traditional music in search of evidence of popular music practices in Colombia in the nineteenth century. If, as a classical work "it is confused with the reality that it describes" (Avelar 2004, 133), this is in part due to the role of ethnographic passages that depict popular practices within the novel. What we see in this early work are the contours of Isaacs's emergent anthropological interests. This Romantic novel depicts the idyllic and unfulfilled love between María, a Jewish convert and devout Catholic, and her cousin Efraín in the slave society of the state of El Cauca. They grew up in the same household because she was adopted by his parents upon her parents' death. He is sent to study medicine abroad, and María becomes ill and dies before Efraín is able to see her again, despite a heroic return trip. Multiple scenes in the novel depict the songs and dances of slaves in the area around the slave-holding estate where the novel takes place, and Efraín's return trip from Europe serves as a narrative backdrop for describing different expressive practices of Afrodescendants in the Pacific Coast.

The novel takes place in Isaacs's native state of El Cauca, a large state that included the Eastern Andean region, politically and geographically subdivided into the southern province of Pasto and the northern patrician one of Popayán, and the Pacific Coast. The region is one of dense tropical jungle, historically the site of intense gold-mining exploitation in which "race became a central organizing factor in the process of settlement and the region's economy" (Escobar 2008, 47). As stated by Escobar, "the view of the region as a sort of pantry or cornucopia of riches to be extracted was inextricably linked to the harnessing of black labor (from colonial slave mining to today's African palm cultivation), not infrequently through representations of race that depicted blacks in natural terms" (47). It has a south-central subregion that is articulated today with the Eastern Andes, particularly in the relation between the port of Buenaventura and the city of Cali and surrounding areas (where the novel takes place), and a northern region, historically more related to Panama and Antioquia via the Atrato River. All of these subregions have significant presence of indigenous and Afrodescendant populations, including the Nasa (formerly Paeces) around patrician Popayán, even though national history has depicted Afrodescendants as restricted to the Pacific Coast and indigenous groups to the southeastern Andean region (Almario 2007). This subdivision into an Andean "more civilized" region and the more "barbarous" jungle area of the Pacific Coast mark the division, in the novel, between different types of African-derived musical practices, producing a "racialized geography" (Escobar 2008) of song.

The bambuco appears repeatedly as sung and danced by African slaves in the estate where the novel takes place. Isaacs's mention of the bambuco as African-derived acquired almost mythical status in nineteenth- and twentieth-century traditional music studies in Colombia as it became the object of repeated etymological debates regarding the European versus African origin of the genre.[28] It appears in the narration of the love story between two African slaves, Nay and Sinar, who come from African monarchic families of different origin. Sinar, who is himself a captive and part of the spoils of war of the "Achanti" tribe to which Nay belongs, would teach her "the dances of his native land, the amorous and sentimental songs (*cantares*) of his native land of Bambuk" ([1867] 2005, 208). Thus, the bambuco appears as African derived but as part of a domestic and domesticated practice characteristic of the Andean region, sung and danced by household slaves of aristocratic origin. Other popular music practices of the Cauca state appear torn between a civilizing and barbarous ideology, depending on whether they are performed in the jungle areas near the coast or in the Andean region.

The most extensive scene depicting such musics is toward the end of the novel when the protagonist, Efraín, is returning from Europe because his beloved María is dying. In order to reach his home, he has to navigate down the Dagua River, which connected the ports in the Pacific Coast with the inland Andean area of the Cauca state. He is taken by two bogas, Laureán and Cortico, who sing a "savage and sentimental song," a *bunde* from the Pacific Coast (Isaacs [1867] 2005, 307). After singing the song, the bogas have a discussion on popular music genres in the region, amidst the sounds of nature that surround them. In the fragment leading to their discussion Isaacs links the descriptions of an acoustic geography of nature to that of the bogas' singing, generating his own notion of "natural music":

> Such singing harmonized painfully with the nature that surrounded us: the late echoes of those immense jungles repeated their profound, slow and plaintive accents.
>
> —No more *bunde*, I told the blacks, in the midst of a pause.
>
> —Do you (*su mercé*) think it is badly sung? Asked Gregorio who was the most communicative.
>
> —No, very sad.
>
> —The *juga*?
>
> —Whatever it is.
>
> —*Alabao*! If when they sing a *juga* well and Mariugenia dances it with this black . . . believe me, sir (*su mercé*) when I tell you, even the angels in heaven move their feet with desire to dance it. (Isaacs [1867] 2005, 307–8)

This passage, more than any other, anticipates María's imminent death (the reason for Efraíns's hurried voyage). The genres mentioned are sung at burials of both adults (*alabaos*) and children (*jugas*) by Afrodescendant communities in the Colombian Pacific, a fact Isaacs obviously knew. Also, jugas are danced and are supposed to have a "happy" feeling, despite being sung at burials because of the belief that children do not suffer in the afterlife but become little angels (*angelitos*) who go straight to heaven. Burial songs sung by slaves also appear earlier in the novel, when Nay (renamed Feliciana in America) dies. The scene mourning her death is a description of the burial practices of Afrodescendants in the region. In the novel, the mourning song is initiated by a solo female slave singer and is responsorially sung by both men and women. Isaacs then transcribes the lyrics to a "hymn" which is, in reality, an alabao, a mourning song for adults by Afrodescendants in the Pacific Coast. But that mourning scene, sung by household slaves, has already been domesticated. The day after the night of mourning, eight slaves and Efraín take Feliciana's body in silence to the burial grounds: "None of those of us who accompanied Feliciana pronounced a single word during the trip. The peasants who caught up with us while taking food to the market found such silence strange, since it is the custom of the villagers of the country to give themselves to a repugnant orgy in the nights that they call the wake, nights in which the relatives and neighbors of the person who has died, get together in the house of those mourning with the pretext of praying for the dead one" (Isaacs [1867] 2005, 235–36).

Burials in the Pacific Coast (including the broader region of El Cauca), effectively take place, even today in some places, not only amidst responsorial songs sung all night but accompanied by food and alcoholic beverages. This is followed, a few days later, by *novenas*, nine-day wakes with song, liquor, and food and, if the dead person is a child, dance. In the novel, whereas the "savage" song sung by the boga takes place in the jungle, the Afrodescendant practices of burial in the estate have been domesticated into a Catholic morale of silence and containment.

This musical geographical division between "civilized" Afrodescendant musical practices from one region and "savage" ones from another is also depicted as autobiographically experienced by Isaacs in the period while he wrote the novel *María*. Isaacs wrote most of *María* during 1864 and 1865 while working as an inspector for the construction of the road (*camino de herradura*) that was being built between Cali (in the Eastern Andes) and Buenaventura, today the largest port city on the Pacific Coast. He was stationed at an encampment called La Víbora in the jungles of the Dagua (Rueda Enciso 2009). He recalled this period as one of the most significant in his life in a letter to his friend Adriano Páez:

There is a period of titanic struggle in my life, between 1864 and 1865. I lived as road inspector on the road of Buenaventura that was beginning to be built in the virgin and unhealthy deserts of the Pacific Coast. I lived then like a savage, at the mercy of the rain, surrounded always by a beautiful nature that was resistant to all civilization, and armed with all the venomous reptiles and all the poisonous exhalations of the jungle. The 300 or 400 black workers (*obreros*) I had under my orders and with whom I lived as a comrade (compañero) had almost an adoration for me. I worked and struggled until I almost fell dead due to the hard work and the bad climate.[29] After that, I have done all my strength permitted, until the Congress of 1878, to work in favor of the road toward redemption of the Cauca region. But that has been nothing compared with what I did and suffered as road inspector from November 1864, to the same month in 1865. (cited by Rueda Enciso 2009, 29)

Isaacs is poised between the civilizatory ideal and his willingness to live "like a savage," a classical conundrum for a citizen-scholar from the postcolony: how to think about commoners while participating actively in the nation's political reforms and simultaneously documenting popular expressions.

The Gran Cauca was characterized by a patriarchal, slave-holding aristocratic elite of lettered landowners who also instituted a colonial regime of servitude of indigenous peoples of the region. The copresence of a powerful aristocratic patriotic elite, and large populations of slaves and free blacks, free peoples of all colors and indigenous peoples, made it the epicenter of struggles between the secular-minded radical liberals and the conservatives.[30] While the radical liberals in power in the mid-nineteenth century promoted the abolition of slavery, they simultaneously "invisibilized" the blacks in the southeastern Andean part of the state by effectively ceasing to name them as existent in the official representations of the region (Almario 2007). Meanwhile the "savageness" of blacks was highlighted in discourses associated with the depiction of their culture in the Pacific Coast. They collapsed the barbarous vegetation of the Pacific Coast into the barbarity of uncivilized Afrodescendants (Almario 2007, 220). As stated by Almario: "With the extinction of the institution of slavery, the black disappeared as a social problem because the reformers thought that they had opened the doors of the kingdom of equality and justice with such measures. But evidence shows the unreality and ambiguity of these discourses that, on the one hand, 'included' in a strange way and without 'naming' them, the blacks from the interior Andean region and that, on the other, definitively marginalized those from the Southern Pacific, but that in doing so, were forced to name them" (2007, 219–20).

María is a novel that has been analyzed from many perspectives.[31] In the last few years, amidst the intensification of work on the history and anthropology of Creoles, Afrodescendants, and indigenous peoples in Colombia, the question of their presence in Isaacs's novel has become central. He has been read as "ambiguous" regarding his position vis à vis Jews, women, and Afrodescendants (Von der Walde 2007), and as ambivalent regarding the question of Afrodescendants in the Pacific (Almario 2007; Múnera 2006). While some see him as condemning slavery and exalting Afrodescendants and others as condoning it, most see his stance as marked by an indecipherable ambivalence. Ultimately what such ambiguity brings forth is a counterpoint of silences and "namings" that has surrounded not only Isaacs's history of publication and the radical liberals' politics in the Cauca state but also the novel and its history of interpretation (Avelar 2004).

As explained by Almario the passage of the bogas of the Dagua River, cited above, has been read by Helcías Matán Góngora, a twentieth-century Afrodescendant poet from the Pacific Coast town Guapi, that is today the epicenter of effervescence of Afrodescendant musical expressions from the Pacific, as foundational not only for Afro-Colombian poetics but for the Négritude movement in general: "Let us not even mention, since it is so well known, the birth certificate of négritude (*Permítame que calle, por sabida, el acta de nacimiento de la negritud*). That is, the manifesto released in Paris by Leighor, Cessaire and Cenaltus. Our birth certificate, in America, was signed, on the tip of the oars of the bogas of the Dagua River, on a page of Jorge Isaacs's *María*. *La Canción del Boga Ausente* from the Colombian poet Candelario Obeso is also earlier in time and geographical space to the Afro-Antillean poetry of Guillén, Ballagas, Palés Mattos [*sic*] and Manuel del Cabral" (Martán Góngora 1978, cited by Almario 2007, 229).

Martán Góngora's play on words—the need not even to mention such an exceptional reinterpretation of the history of Négritude—rests on the tension between the silenced (and the unrecognized) yet named and known, as the main interpretive trope of (Colombian) négritude in the Cauca region. He claims that Isaacs and Obeso are foundational to Afro-Antillean poetry by invoking a "politics of the prior" (Povinelli 2011): they appear earlier in history and in the American geographical space than any of the other Antillean poets he mentions, yet as an origin that has not been recognized. Such a movement between revelatory silences and foundational historical claims bring to the foreground the novel's use as a resource for knowledge about Afrodescendant expressive culture: "Most interpretations that attempt to trace *María* back to its historical, cultural soil struggle against the same paradox: They refer to a history and a culture that have been considerably shaped by the novel's fabulation itself. This is, then, a classic: a text that, facing the critic's attempt to circumscribe its conditions of possibility, refers back at him/her a false base, a product of its own game of mirrors (Avelar 2004, 133).

In the case of *María*, such a game of mirrors stands as much on the difficulties of establishing a clear dividing line between fiction and nineteenth-century ethnography that characterizes the novel itself, as on the concatenation of the specters of multiple silencings, not only in the novel, but in its interpretive history. As Avelar himself masterfully unveils, *María's* "politico-ideological operation" takes place in "the terrain of the unsaid" (2004, 136). For Avelar, such politics of the unsaid rest on the differential treatment of Jewishness and Afrodescendants in the novel. While, according to Avelar, Afrodescendants are brought into the realm of silencings characteristic of cordiality, the silencing of Jewishness is structured through a different process, a "politics of the unsaid" that Avelar uncovers gradually through a buildup of a history of multiple terms of negation that structure his critique.

First, we have the "*early forgetting*" and "*erasure*" of María's Jewishness, initiated (no less and not accidentally in light of the earlier part of this chapter) by Vergara y Vergara's prologue to the second edition of the novel. This "inaugurates a long tradition of *denial* of Jewish heritage in the text," an erasure that becomes "*a constitutive moment in the text's history*" (emphasis in the original) (Avelar 2004, 137). "*Erasure*," says Avelar, "is here understood in the Derridean sense of an operation that at the same time *hides and shows* that it is hiding" (137), constituting a spectral presence. This is then linked by Avelar to notions of Freudian and Marxian *negation*: "what is negated is always and necessarily making possible negation itself. As you negate, a fundamental *turned around* (italics in the original) form of affirmation emerges" (138) that ultimately appears as a *spectre* that "comes to haunt the act of negation itself," particularly in the way that "all the critics that have devoted time to negating the 'Hebrew' or 'Jewish' 'influence' in *María* seem to become, involuntarily, characters of what the novel describes. They all seem to be transported to the scene depicted in the text. In being engulfed by the text's elastic throat, they thereby lose the possibility of reading it critically, for negation and masking are the text's very themes" (138). Avelar then explains in detail the novel's fundamental allegory. María is born Esther daughter of Sara and Salomon and, upon the death of her mother, given by her father to be raised by Christian relatives, renamed as María. Thus, "the name *abandoned* by María at baptism confers on Isaacs's allegory its definitive meaning" (138). Avelar's elucidation of the multiple modes of denial reveal the different concatenations of silencings at work in the novel, in Isaacs's life, and in its successive historical interpretations.

The fundamental allegory that Isaacs is portraying through this concatenation of silencings is that of the centrality of Judaism in defining the idea of race in Colombia as one of negations within negations. In eighteenth-century New Granada, the demonstration of blood purity was the fundamental *legal* operation for having access to the lettered city since, without it, it was not possible to receive a university education or to hold a position in the administrative institutions of the state or to get legally married (Castro-Gómez 2004; Silva 2005). Moreover, the tribunal of the Inquisition in Cartagena which operated from the sixteenth century until the moment of independence, roughly sixty years prior to the publication of Isaacs's novel, was as concerned about *conversos*, Jews or Muslims hidden under the cover of Christianity, as about practices of Afrodescendant wizardry and indigenous shamanism (Borja Gómez 1996). The basic operation of survival of the converso is that of a silencing through renaming, the trope of Isaacs's novel. Isaacs, the son of a Jamaican Jew who married a Catholic woman was ultimately, for the lettered city, always a converso, one who questioned its Catholicism through his haunting presence in the nationalist nineteenth-century Catholic Hispanist project.[32] In Colombia, as in Mexico, the expression to "speak in Christian" (*hable en cristiano*) is used, even today, to mean speak Spanish well, an expression that transfers the politics of *bien decir* to the spectral history of silencing and naming that the racial/religious politics against conversos hide (Lomnitz 2001). If critics confuse the novel with the author and with the author's place in the lettered city, it is also because it depicts the fundamental mark of Isaacs's life. Also the similarity of this operation of fictionalization with the politics of radical liberals in the Cauca described by Almario as regarding Afrodescendants, that is, ceasing to name them as existent once slavery was abolished, or naming them but confining them to what were considered the most backward outskirts of the nation, is significant. Isaacs moved between a multiplicity of techniques of silencing, masking and (re)naming casting both his figure and the novel into a history of reinterpretations of negation and naming as the basic operation of violence he enacts (Avelar 2004) and with which he was also marked.

In his later life Isaacs became one of the most radical of the radical liberals, devoted to the cause of secular education as superintendent of public education of the state of Cauca when his cousin, poet and philologist César Conto, became president of the state of Cauca between 1855 and 1857. As we will see in the next chapter, toward the end of his life, his radical politics would increasingly mark his texts as Isaacs became more and more convinced of a politics of listening that would shape his ethnographic stance, becoming less the cordial citizen he was in the early years of *María*. He would end his life ostracized by the alliance of the conservative elites with the more moderate liberals that ultimately led to the triumph of the Hispanist Catholic conservatives in the 1880s. But in this early phase of his career, *María* emerges as an emblem of the contradictions of documentation of the aural in the written word that would characterize the emergence of musicology in Latin America in the hands of literary figures in the early twentieth century: that of a purposeful blurring of the dividing line between ethnography and fiction in the recognition of the debt of the uniqueness of the American written word to its aural inscription, as a conscious impulse of the avant-garde (Rama [1984] 2007; Ramos [1989] 2003). This is as much a history of anthropology as one of literature and musicology, since the novel as ethnography permitted the reflexive stance demanded of an anthropology of music for one that is a member of the same polis. It is, of course, overtly steeped in the ideological work of such figures, since its pretension is not to describe an outside but to participate in the redefinition of the diversity of which the writer is a part.

Conclusion: Songbook Inscription between Recognition and Silence

Vergara y Vergara proposed a project that became programmatic by the early twentieth century as books on coplas following his model became one of the main products of folkloristics. Folkloristics carried the pretension of documentation mediated by the history of a racialized geopolitics of documentation here depicted in Vergara y Vergara's work. But as we saw in this chapter, different authors propose a different relation between audile techniques and a politics of inscription. Ultimately, when analyzing comparatively the work of the three authors explored here, one sees that the history of popular song, in its politics of inscription, reveals as much about acoustic practices as about their concealment, and is thus simultaneously a history of the inscription of song as of the silencing that such inscription necessarily produces. Such silencing speaks to the politics of enunciation of each author, to the spectral relation between that which is acknowledged and that which is negated, as well as to the type of materiality that mishearing, produces on the page. The gap between auditory perception, auditory sensation, and inscription emerges in the relation between silence, mishearing, and misspelling that produces the idea of "orality" in song.

The obsession with blood purity of New Granadian eighteenth-century elites spoke not only to the impossibility of keeping apart the intense process of mixture of races that was happening in Colombia in the eighteenth century (particularly in the Caribbean and Cauca, those two places with many *libres de todos los colores*) but also the difficulty of assigning the mechanism of blood purity for racial differentiation at the moment of the postcolony. The documentation of "popular poetry" or folk song as a central element of nation-building implied a process of silencing and recognition that was, in all cases we explored in this chapter, transversed by the relation between race and religion. Along with other philological and political developments of the nineteenth century we will explore in the following chapters, what we see here is the imbrication of religion and culture for the nation, in the "public need to renew and maintain identifications and affiliations, to preserve existing structures by preventing confusion" (Anidjar 2008, 17). As we saw throughout this chapter, this is a history of the way "two disciplinary—albeit very different—marks, namely, religion and race" are enmeshed together, "and the strategic uses whereby one obfuscates the other" (17). If for Afrodescendants, the basic operation of the transformation of the idea of race in the initial years of the postcolony was from blood purity to soil in the tension between a musical voice and the domestication of *mal-decir*, for Jewishness it was that which persists as hidden (blood) in the haunting of a language that silences as it (re) names, and for indigenous peoples that of a silence that perpetuates extermination by anticipating the silence of death as already accomplished and therefore forever repeatable as silenced/t beings. The history of inscription of song then is as much a history of its heterogeneous acoustics as of a concatenation of silencings, a history of the secularization and racialization of the voice as a critical component of the idea of orality.

Nineteenth-century Latin American philologists were acutely aware of the disjunction between vocal pronunciation and alphabetic writing as the central problem presented by the science of orthography. Take, for example, Colombian philologist Ezequiel Uricoechea's (1834–1880) disquisition on such a topic in the introduction to his edition of the *Gramática, vocabulario, catecismo i confesionario de la lengua Chibcha según antiguos manuscritos anónimos e inéditos, anotados y correjidos por E. Uricoechea* (Grammar, vocabulary, catechism and confessionary of the Chibcha language according to ancient anonymous and unedited manuscripts, annotated and corrected by E. Uricoechea) (1871), the inaugural tome of his *Collection Linguistique Américaine*:

> If orthography is nothing more than the art of good writing, and this art consists of representing articulated sounds graphically and exactly, reserving one sign for each sound, there is no doubt that no orthography today fulfills its object. More than eighty different sounds are known in the living tongues and even though there is no tongue that possesses them all, neither is there an alphabet that represents its own. To take this absurd situation to the extreme, through the conservation of signs from other tongues, we often have two or three signs that represent the same sound. . . . The preconceived idea that an orthographic sign represents the same sound in all the tongues, and, if I may say, the universal and absolute lack of knowledge on the pronunciation of the alphabet in each one of them, because no one has yet studied each in depth, is the cause of the difficulty in learning foreign living languages and

of the fact that it is almost impossible to destroy the accent peculiar to each nation. . . . If we recognize these difficulties today and we feel the inadequacy of our alphabets, how much more difficult must be the knowledge of the pronunciation of tongues that have been lost for a century when orthology was not yet known and orthography was the patrimony of a few? (1871, xlvii–xlviii)

Uricoechea's particular concern is directed toward the difficulties he encounters while editing for publication in the collection he founded, the *Collection Linguistique Américaine*, the indigenous grammars compiled by missionaries in New Granada between the sixteenth and nineteenth centuries. His interest in this project lay in his desire to undertake a proper systematization of the grammars of indigenous languages in the Americas in order to compare them to Asian languages and thereby answer questions about the origins and nature of the American continent and its peoples (Uricoechea 1871).[1] In such a project the history of the compilation of missionary grammars in Latin America from the sixteenth to the nineteenth centuries came together with comparative philology's quest for an original Indo-European language.

Both the American missionary project of indigenous grammars and German comparative philology were politico-theological enterprises centered on metaphysical questions. The first, a political project of conversion, rested on questions regarding the "soul" and "practices of idolatry" of indigenous inhabitants and on whether America was a demonic or a paradisiacal land, in short, on questions regarding the nature of the American continent and its peoples (Bernard and Gruzinski 1988). The second was spurred by Sir William Jones's (1746–1794) "discovery" of the linguistic affinities between Sanskrit, Greek, and Latin, the consequent establishment of the comparative grammar enterprise, and the search for a common Indo-European ancestral language (Benes 2008). This would supposedly yield crucial understandings of bygone eras, including a response to the questions of whether humans were derived from one single Adamic race and the location of paradise (Olender 1992; Benes 2008). Although such questions were not new either in the Americas or in Europe, they took new form through the method provided by nineteenth-century German philology of comparing either sound/lexical affinities or grammatical relations between words. This interest in origins led comparative philologists to interpret "language as evidence of ethnic descent" (Benes 2008, 10). With these interpretations they "wove myths of cultural origin around the perceived starting points of national tongues" (10). In the Americas this historical principle of comparative philology was brought to bear on questions regarding the perceived origins of the continent and the nature of its inhabitants.

Both Ezequiel Uricoechea and Jorge Isaacs theoretically reframed the historical missionary grammars and sought to compile new grammars and vocabularies of Colombian indigenous languages. These languages' validity was judged either academically by the keys they provided about the history of the American continent (Uricoechea and Isaacs), or aesthetically, with their beauty making them worthy of respect and study for their own sake (Isaacs). By recasting indigenous grammars as historical "data" about the origins of the American continent or simply as contemporary linguistically valid tongues, the Colombian intellectuals who studied these languages in the nineteenth century either set the stage for opposing (Ezequiel Uricoechea) or overtly opposed (Jorge Isaacs) the political-theological authority of the conquest and of the nascent postcolony. In doing so, they challenged the mode of narrating the history of the American continent, questioned the authority of the grammarians by considering new modes of understanding the relation between the sound of words and their inscription, and problematized the place of indigenous peoples in the new nations.

Uricoechea's cosmopolitan project was developed in Europe through a passionate diplomacy that sought to frame a decolonial Americanist project by searching for the origins of the American continent and of its peoples. Far from the political obfuscations of his homeland, he managed to evade the pettiness of everyday local politics and mediate between multiple Colombian philologists through correspondence and an active trade in books between friends in Europe and the Americas. In contrast, when Isaacs wrote his *Estudio sobre las tribus indígenas del Magdalena* (Study of the Indigenous Tribes of the Magdalena Region) in 1884 he had politically questioned the grammarian state as a soldier, educator, and writer. He overtly challenged the religious and political motives for which such indigenous grammars were, even then, being studied by missionaries. Such a challenge was not lost on Miguel Antonio Caro (1843–1909), soon to be vice president of the republic, who was then busy authoring the nation's new constitution. He responded to Isaacs's proposal of studying the languages firsthand by living among the indigenous groups "in order to learn from them" with a vituperous fifty-page critique entitled "El Darwinismo y las misiones" (Darwinism and the Missions) (1886). In principle, what Miguel Antonio Caro found preposterous was precisely the transformation of the politico-theological authority of such a scholarly enterprise:

> From the scientific point of view, the philology of barbaric languages owes all of its wealth to the missionaries. The interest in studying those

languages has always been subordinated to teaching the Christian doctrine to those who speak them. Without this religious motivation, who
would have bothered to study seriously the jargon of the savages? Who,
in studying grammars and vocabularies? Nor who would have thought
of establishing, as did the Jesuits in Bogotá, classes on such unliterary
languages? Would it not have been more natural and easier to despise
them highly and thus contribute to their extinction, persecuting the
very same tribes that speak them? The condescension of studying the
speech of the savages was not, has not been and will never be the effect
of a natural impulse but the work of grace. Nobody is going to relate to
savages for love of the science of language; no philologist visits the barbaric tribes for the pleasure of taking materials first hand. (Caro [1886]
1980, 1090)

In this chapter I explore how attention to indigenous word sounds and the
effort of rethinking their orthographic inscription in the nineteenth century
gave rise to a historical and ethnographic ear that either potentially or overtly
challenged the politico-theological authority of the conquest and of the governing Creole elite's postcolonial rearticulation of indigenous issues in the
nation. Such a question will be explored initially through philologist Ezequiel
Uricoechea's work, particularly his theories about the origin of the American
continent and its peoples and the way such questions articulated ideologies
of language and language sound in his edition of New Granadian missionary
grammars. I will also discuss the controversy regarding the place of indigenous
peoples in the nation provoked by one of New Granada's earliest ethnographies, Jorge Isaacs's *Estudio sobre las tribus indígenas del Magdalena* (1884), and
philologist Miguel Antonio Caro's response to this text, "El Darwinismo y las
misiones" (1886).

In challenging political authority, Uricoechea and Isaacs opened the way
for a profound reevaluation of the significance of the American continent
through its linguistic history. Debate on the relationship between indigenous
languages and the origins of the continent gave rise to a theory of difference
couched in the disjuncture between the sounds of indigenous languages and
the mode of rendering them through orthography. If, in Saussure, the problematic materiality of language generated an ontological doubt as to its status
as an *object* of science, evident in the disjuncture between signifier and signified and between the physiological articulation of vocal sounds and their
sonorous rendition (Maniglier 2006) in the Americas, the impossibility of a
proper transcription of an indigenous tongue yielded an ontological doubt as

to the status of the American continent. Such a "linguistic turn," peculiar to the region, was articulated in the nineteenth century through the impossibility of properly transcribing the continent's linguistic-acoustic marks, made evident by the gap between indigenous linguistic sounds and alphabetic graphemes. If the lettered city was built around the authority of a "priestly caste" of lettered men (Rama [1984] 1996) who defined the form of political authority in Latin America, the disjuncture between indigenous linguistic sounds and their inscription challenged the place of such authority in deciphering the region's ontology.

Thus emerges narration's profound philosophical status in Latin America, in which the motives of a speculative realism (regarding the nature of the region) and those of language are not opposed (as in the idea of a linguistic versus a realist "turn")[2] but rather entwined in the decipherment of the given. Here the question of language was ambiguously mobilized for different philosophical purposes. On the one hand, in using it to define the nature of the continent, "language, as a limited tool, is an empirical part of the world to which it refers, rather than a transcendental condition of that world" (Shaviro 2009, 150). Yet debates on the value of indigenous languages also provoked the invocation of Spanish as the transcendental condition for the theologico-political definition of citizenship. Rendering invalid indigenous forms of personhood and speech territorially and philosophically circumscribed the phantasm of originary America that the contemporary presence of indigenous peoples posed for the new nations. This chapter is about the ontological questions raised by a passionate Romantic investment that sought to rethink the nature of America through its indigenous languages, the politico-theological background against which those questions were formulated and the backlash they generated, a debate centered on the disjuncture between hearing and inscribing indigenous tongues as central to the emergence of ethnographic thinking. At stake was the juridico-political status of indigenous persons in the emerging nation.

Ezequiel Uricoechea and the Question of American Origins

"*Esta mi manía de cosas americanas es insaciable*" (This, my mania for American things is insatiable) wrote Ezequiel Uricoechea from Paris to his friend Juan María Gutiérrez in Buenos Aires (Uricoechea 1998, 102). Uricoechea was probably the one with the most cosmopolitan vocation of the several internationally renowned philologists of mid to late nineteenth-century Colombia. An orphan at an early age, he was sent by his older brother to finish high school in the Flushing Institute in Long Island, then graduated from Yale with a degree in medicine at the precocious age of eighteen.[3] From there he went to Gottingen "through the recommendation of Humboldt" (Uricoechea 1998, 294) where he obtained a doctoral degree with a dissertation on chemistry and mineralogy in 1854. That same year he published his first book, *Memoria sobre las antigüedades neo-granadinas* (Memoir on New Granadian antiquities), initiating a lifelong production of works that would move between the natural sciences, antiquarianism, and philology. From Gottingen he went to Brussels, where he studied astronomy and meteorology, and finally returned to Colombia in 1857.

In Bogotá he taught chemistry in the Colegio Mayor de Nuestra Señora del Rosario for ten years, founded the Sociedad de Naturalistas Neogranadinos (Society of New Granadian Naturalists), directed the society's publication of *Contribuciones de Colombia a las ciencias i las artes* (Colombian Contribution to the Sciences and the Arts [1859–61]), and became a member of El Mosaico. But due to the instability generated by civil wars, and to repeated frustrations with several projects, Uricoechea returned to Paris in 1869[4] where he dedicated himself to philological studies and became an avid Americanist in Europe. He also studied Arabic, purportedly "to learn the Arabic terms used in mineralogy and metallurgy for his edition of his *Tratado de mineralogía* (Mineralogy Treatise)" (Botero 2002, 27). In July 1878 he became the first professor of Arabic at the Free University of Brussels. He is thus known as "the first Orientalist from Colombia" (Hernández de Alba 1968). In 1880 he began a trip to Damascus via Alexandria and Beirut, where he had planned, as stated in his last letter to Colombian philologist Rufino José Cuervo, "to rent a little house, take a cook and a servant . . . and, if possible, find a female teacher (*una maestra*) of Arabic, because the teacher with whom I wish to consult my grammar of vernacular Arabic, the one I showed you, I will find on the street. If I manage to install myself thus, in family, I will stay there all the time, except for one month during which I will go to live with a tribe. Voilà mon plan" (Uricoechea 1976, 258). Uricoechea died "of a fulminating disease" in Beirut on July 28, 1880, en route to his Orientalist dream. He is considered along with Miguel Antonio Caro and Rufino José Cuervo one of the foremost Colombian philologists (Hernández de Alba 1968).

His cosmopolitanism undoubtedly derives from his transnational life. It manifested most strongly in a cartographic imagination that took form in an obsession to collect in order to create a map of Americanist thought and in his multiple and repeated efforts to disseminate the work of Latin American scholars in Europe and between American nations. He thought to unsettle the epistemic dominance of Europe, in a decolonial project framed by a desire to construct a "Latin Americanism" (De la Campa 1999) that would become the true sign of independence of the region. In a letter to Juan María Gutiérrez he wrote: "We are in such state of dependency, we have become such slaves, that none but them possess science and, if we have gods, only the European ones are venerated on our altars. We will only be independent the day we are in body and soul" (Uricoechea 1998, 110, letter to Juan María Gutiérrez, April 1, 1872).

To achieve such a project he became an expert compiler and disseminator of information. He compiled a historically organized list of titles of maps of Latin America in his *Mapoteca colombiana, colección de los títulos de todos los mapas, planos, vistas, etc. relativos a la América Española, Brasil e Islas adyacentes* (Trubner: London, 1860), and a bibliography on Colombia, *Bibliografía colombiana*, which contained 4,000 titles, only partially published (Botero 2002, 14); he wrote a dictionary of natural history terms, *Diccionario de voces de historia natural americanas* (1873), "which involved the translation into scientific language of more than five thousand vulgate names of American species" (Botero 2002, 23), a project that also remained unpublished despite his efforts, and that was lost upon his death. He began to compile, edit, and publish the Amerindian grammars written by missionaries in the Americas, beginning with those of New Granada in the *Collection Linguistique Américaine* (later *Bibliothèque Linguistique Américaine*). He was able to edit three volumes before he died, but the collection continued after his death to include a total of twenty-five volumes. Besides other compiled works, he also wrote an *Alfabético fonético de la lengua castellana* (1872), written as a letter addressed to Spanish philologist Juan Eugenio Hartzenbusch. Except for his early *Antigüedades neo-granadinas*, his oeuvre consists of comparative lists, introductions to his own and other peoples' works, and a disquisition on phonetics written in the form of a letter. These are highly intertextual genres that seek to establish relations between things and people and between different people.

A "maniac" about things American yet unable, by cosmopolitan vocation, to live in his own country, Uricoechea created, in his innumerable lists and introductions, the poetics and politics of a relational Americanist cartography. By distancing himself from the civil wars and partisan struggles of the period, he allowed for questions that perhaps would not have flourished otherwise. For while his colleagues in Colombia were obsessed with creating the epistemological moves that were crucial for binding knowledge to the economic labor of the political in New Granada by mapping nature to order the nation (Nieto Olarte 2007), or the songs to represent it (chapter 2), or the perfect rhetoric with which to speak it (Von der Walde 1997; Rodríguez-García 2010), Uricoechea sought to grasp the portent of it all through an avid passion. Removed from the daily routines of local political struggles, his obsession became the continent rather than the nation or, as he says, "things American." One could say that he had, in the sense of its amplitude, a Humboldtian imagination, yet one that was attuned to a decolonial politics of recognition of America's place in history and in his contemporary world. In such passionate commitment to America, he lets himself be positively overtaken by the "inexhaustible multiplicity that envelops him" (Zalamea 2009, 25). Uricoechea seems to be constantly giving himself to his words, his objects, his maps, and his interlocutors; to those *things* he so avidly collects and the people and worlds he so much wants to link. His constant use of a hyperbolic and emotionally expressive language denotes not an exaggeration but a "becoming undone" (Grosz 2011) in his project. His is the language of passionate love. As he himself says in a letter to Juan María Gutiérrez:

> I have not been able to stop being an American even though I have often desired it in times of revolt. But an American in the full extension of the word, from my bone marrow to my skin, from the cradle . . . to the tomb, which I will not be able to avoid. And all that comes from there, when it is good, is a balsam that not even Cagliostro had, and that no alchemist ever even imagined. . . . Do not be surprised then at my outbursts. I am happy. And if we excuse those in love many times . . . and even envy them, I believe others also deserve indulgence when the heart speaks. (Madrid, August 29, 1872, in Uricoechea 1998, 120)

Such joy in the production of knowledge manifested itself as a desire to bring into relation "all that comes from there" in order to understand what America names. In his introduction to the *Gramática, vocabulario, catecismo i confesionario de la lengua Chibcha*, the first tome of the *Collection Linguistique Américaine*, he leaves no doubt as to the magnitude of his project: he wants to complete the labor begun by Jesuit linguist Lorenzo Hervás y Panduro (1735–1809) and continued by Adelung and Vater. Besides the pioneering efforts of these scholars, says Uricoechea, "no one has yet undertaken the comparative study of American tongues amongst them and with the Asiatic ones" (Uricoechea 1871, xiii–xiv). He thus sees himself as a person mediating between German philology and the work of Hervás y Panduro through the study of Amerindian grammars.

The rise of German comparative philology in the eighteenth and nineteenth centuries "made the interpretation of words central to a historical definition of cultures and to an ethnological project of establishing genealogical relations among the world's nations" (Benes 2004, 118). Such "Mosaic ethnology" (Benes 2004, 2008) arose from a biblical concern of accounting for human origins as told in Genesis (Adamic descent), for historically tracing the dispersal of peoples throughout the globe as descendants of Shem, Ham, and Japheth, Noah's three sons, and for identifying the relation between the dispersed tongues and peoples of Babel (Benes 2004, 2008). In such a project etymology became a means to "uncover how prehistoric peoples had lived, thought and worshipped" (2008, 15) and such linguistic research eventually produced a transformation from biblical and Christian terms to historical ones in the definition of cultural difference.

This gave rise, in the eighteenth century, to large collections of comparative linguistic items of languages collected through the colonial missionary enterprise across the globe.[5] The most famous of these collections in the eighteenth century included those of P. S. Pallas, a collection of 285 words in European and Asiatic languages that was first published in 1787 under the sponsorship of Catherine the Great and reedited in 1790–1791 in order to include information about 80 more languages, among them African and Asian languages;[6] the Adelung-Vater collection called Mithridates (four volumes, 1806–1816), essentially a list of the Lord's Prayer in some 500 languages (Jankowsky 1972); and that of Hervás y Panduro, who published twenty-one volumes of an "encyclopedia" entitled *Idea del l'Universo* (Idea of the Universe) in Cessna, Italy between 1778 and 1787, the last five volumes (17–21) of which are dedicated to the study of the languages of the world (Tovar 1986).

Don Lorenzo Hervás y Panduro (1735–1809) was a Spanish Jesuit priest and linguist who spent most of his life exiled in Italy due to the expulsion of all Jesuits from Spanish domains in 1767. Many of his informants were other Jesuits who had also been forced to return to the papal states and who had compiled grammars, vocabularies, and comparable texts to be used in the missionizing enterprise in multiple languages around the world, including the Americas. Hervás y Panduro wanted to "compile a history of languages of the world, by which we are to understand a descriptive study of their […] pronunciation, morphology, syntax and vocabulary, for the purposes of cataloguing, setting up family groups and revising the history of nations" (Breva-Claramonte 2001, 266–67). Spanish linguists today point out his crucial historical role: he was probably one of the men of his time most thoroughly informed about languages around the world, and his collection of languages was vital to the formation of the Adelung-Vater Mithridates and for Wilhelm von Humboldt's knowledge of languages in the Americas.[7]

These collections are usually considered part of a prescientific philological discipline that was mired in fantastic comparisons across peoples and languages (Davies 1998). Around the 1830s there is a transformation in German philology from great collections of language comparison, in which linguistic facts were meant to help address historical questions, to interest in the study of linguistic aspects of language per se. This eventually led to the separation of linguistics from history and ethnology (Davies 1998). And despite the evaluation of eighteenth-century philological work as "prescientific," the comparative theorization of the nineteenth century depended, in good measure, on the vast amount of information compiled in that period. However, such transformation did not solely take place in Germany. Rather, it was part of a vast global enterprise of redefinition of the role of linguistics that included the relationship between the history of missionary grammars and European philology (Zwartjes and Hovdhaugen 2004; Zwartjes and Altman 2005; Zwartjes and Ridruejo 2007). In fact, New Granadian philologists were themselves more than aware of their place within the comparative enterprise.[8] But, among them, ultimately it was Ezequiel Uricoechea who played a mediating role between Hervás y Panduro, the new German philological comparative sciences, and Latin American philology.

Uricoechea's interest in salvaging indigenous grammars was directly related to the Orientalist impulse provided by his training in German comparative philology and mediated by his intention to continue Hervas's project by recovering the grammars of missionary priests in New Granada. The extensive introductions to the missionary New Granadian grammars he edited are analytical works in which he uses the data compiled by missionaries to locate indigenous peoples and their languages in a comparative American ethnological map. Ultimately, such linguistic expertise was to be used in answering the question that obsessed him, that of the nature and origins of America.

This question first shows up in his early *Memoria sobre las antigüedades neogranadinas* (1854) where he defends a monogenetic conception of history based on Adamic origins of humanity, "one primordial origin for all humankind and so the question is to know from what trunk or family of the old continent was the new populated or vice versa" (Uricoechea 1854, 6). Such an affirmation is couched amidst several doubts. On the one hand, he acknowledges that the new evolutionary theories that had recently emerged are difficult to accept possibly "because of a certain inner pride" (6) and because the debate is so recent that it is difficult to take sides adequately and assume its theological implications. Uricoechea tacitly recognizes that Darwinian theories are an attack on a Christian and anthropocentric idea of life. They are difficult to accept because to do so would imply acknowledging man's descent from animals challenging notions of identity, of human privilege, bodily boundedness, rational exceptionality, and spiritual superiority (Grosz 2011). In this regard, Darwinism was a "profound insult to man's sense of self" (Grosz 2011, 13). So Uricoechea chooses to defend Adamic theories of creation of man and steers the issue onto more secure grounds by asking a question about place: whether the origins of humanity are in the Americas or in Europe.

Throughout his work and correspondence, these three elements—Adamic origins, the question of evolutionism, and America as an older continent than Europe and the place from which humanity emerged—appear again and again. The first question is genealogically linked to German comparative grammar, the second to Darwin's theories, and the third to theories developed during the colonial era that depicted paradise as located in Andean America.[9] The formulation of how he conceives the relation between these three elements changes throughout his life but guides his whole intellectual quest. Here, his mode of phrasing the issue in the introduction to his *Mapoteca colombiana* (1860), six years after his earliest formulations:

> These men [indigenous peoples] appear in America without any previous mention of them neither in history nor through revelation. Thus, no one had suspected their existence. Several authors have tried in vain to explain their descent from the Asian peoples that populated Europe and their travel to America. Either we have to consider that in former times there was a very different order of things from that which exists today, admitting, in the first place, a difference in configuration of the present-day continents, in order to admit migration in times in which men were so backward in the art of navigation, or we could suppose rather, something that is perhaps more viable, a special creation in America. This theory is in no way opposed to the general one established by geologists, regarding the repeated and simultaneous creations of several species of other organized beings. But since this stands in opposition to several mythological and religious beliefs, it has not had any acceptance regarding man. This point will probably never have a satisfactory solution, despite the most detailed research and all just are afraid of erring by accepting opinions from any writer. (1860, xiii)

Uricoechea was hesitant about fully confronting the proposition of evolutionism because of the impossibility of addressing it without overtly questioning the theological establishment. Ultimately, as happened with German comparative philology, "exposing the origins of metaphysical concepts could . . . fatally unmask the contingent, nonbinding foundations of theology and ethics" (Benes 2008, 11). So while his world was enriched by the labors of comparing the fascinating multiplicity he finds in America, he cannot address what he purports to solve, the singularity of American origins.

In later formulations of this idea he simply embeds the idea of origins in an ambiguous statement. For example, he finds that the similarities in arrowheads, vases, axes, and needles found in tombs in Europe and in the Americas, tools that he finds to be "identical" despite great distances, are "proof of a single origin," that is, "they are the product of a same school and thus of contact between manufacturing peoples. The hypothesis that the oriental and occidental peoples of the worlds were one before the great historical flood or of that of which we have news due to the Old Testament, gains more acceptance every day. Once that is admitted, we could have no better explanation for what we see" (letter to Gutiérrez, December 4, 1873 [1998], 153–55). Couched in either "historical" or "biblical" accounts of the great flood, the idea of "a single origin" of Oriental and Occidental peoples could potentially incorporate either Adamic or Darwinian narratives. Entangled between the religious commitment of Adamic philology and the uncertainty of scientific discoveries that seemed logical but were still to be developed or fully accepted, Uricoechea, rather than challenge either his philologist friends or maybe his own religious convictions, chooses to stand on the brink. It is as if Uricoechea was unable to take the step from "that which is admissible to that which is possible" (Sloterdijk [1988] 2006, 40), or perhaps it was not yet the historical moment to do so. So, Uricoechea formulates an ambiguous theory that allows for both a scientific and a religious reading of a general law of common origins, hesitating to fully assume the metaphysical implications of moving from a theory of history (philology) to a transformative theory of life (Darwinian evolutionism) in the creation of a new ontology for America. Part of the difficulty of making this move also rests on his perception of indigenous peoples. Even though he admits that "they were far from being the barbaric peoples depicted by the chroniclers," they remained "stationary" in their development, and thus, ultimately for him the American continent ended with a more backward civilization than Europe (Uricoechea, letter to Juan María Gutiérrez, December 4, 1873 [1998], 154).

Faced with the daunting realization of the impossible decisions that the search for origins entailed, Uricoechea turned to new beginnings. He became an Orientalist, began to study Arabic, and developed, in the reedition of indigenous grammars, a new poetics. There is no nascence without othering (Arendt [1958] 1998), so it is to stories of othering and to language that Uricoechea turns. Unable to unveil the question of origins of humans in the continent, he turns to anthropology (who were the inhabitants of America?) to answer an ontological question (what is America?). He retakes the colonial chronicles of conquest and the missionary grammars in order to recast the information they contain as ethnographic and linguistic. In such rewriting he begins a process of transmission and transformation of the colonial archive, and of reinscription and resignification of indigenous languages. Here the role of tradition-transmission is to account for (new) beginnings as modes of transformation and reinscription (Sloterdijk [1988] 2006) in a poetics in which the relation between the original and the copy are understood as the means of repetition in the act of creation. It implies an understanding of the emergence of novelty from what already exists (Whitehead [1927/28] 1978). In such a move the crucial epistemological and ontological question changes from one of origins to one of the relation between nature and culture, mediated by the linguistic sign.

Linguistic Disjuncture and the Ontology of the American Continent

Uricoechea edited and published three missionary grammars and vocabularies: *Gramática, vocabulario, catecismo i confesionario de la Lengua Chibcha según antiguos manuscritos anónimos e inéditos, aumentados i correjidos* (1871); *Vocabulario Páez-Castellano, catecismo, nociones gramaticales i dos pláticas conforme a lo que escribió el señor Eujenio del Castillo i Orosco con adiciones, correcciones i un vocabulario castellano-páez por Ezequiel Uricoechea* (1877); *Gramática, catecismo i vocabulario de la lengua Goajira por Rafael Celedón con una introducción i un apéndice por E. Uricoechea* (1878). The first of these has a different beginning than all the rest, one that, in a hyperbolic, celebratory and gendered poetic language depicts the generous magnificence and abundance of American nature and topography for the colonist:

> This continent contains in its womb the most precious stones such as the diamond and emerald, and its open breasts have yielded, for the past three hundred years, rivers of gold and silver over insatiable and thirsty humanity. There we can find in prodigious abundance the iron, copper, lead and platinum that men seek with ardor and that she lovingly gives them. A land of all climates, it gives man a welcome asylum and offers him, amongst the tropics, a constant temperature, always the same throughout the year in every place. In the tropical zone the colonist chooses between the African heat found in the coasts and all the

other temperatures, from the most smooth ones to the perpetual ice found on the mountainsides or mountaintops, whose high plains are the valleys of an Eden that only the poet dreams. In its immense plains we find an infinite number of domestic animals that having emigrated with man, seem to have returned to their own home, such is their prodigious development. . . . (1871, x)

And so on. Upon the failure of the impossible task of proving the existence of humans and of Eden in America, he turns his writing into an all-giving maternal paradise upon which the male colonizer can placidly rest in abundance. The interpretation of Los Andes as Eden, depicted as a region that contained all possible climates and a great number of species, derived from the meeting of Amerindian and European constructions of space (Cañizares–Esguerra 2006, 116). The indigenous practice of a vertical cultivation of different plant species in various niches of altitude was not only mobilized for commercial purposes by the Spaniards but subsumed under paradisiacal readings of the Andes as a biblical Eden, in the ellipsis between indigenous myths of mother earth and Marian cults (116). By the eighteenth century, naturalist José Celestino Mutis (1732–1808) and the fellow members of the botanical expedition of New Granada had mobilized the idea that the Andes contained microclimates capable of producing any plant in the world, thus positing the viceroyalty of New Granada as one with unlimited economic potential. New Granadian naturalist Francisco José de Caldas gave scientific form to these theories, developing a patriotic discourse around the biodistribution of plants (Cañizares-Esguerra 2006). Uricoechea was one of several writers who developed a form of narration that incorporated the description of the features of nature in the continent into a celebration of abundance, a prequel to the identification of the marvelous real as the mark of the continent.

Uricoechea used such a narrative to preface ethnographic descriptions of indigenous groups and their languages in the introductions to the indigenous grammars he edited. He linked the hyperabundance of nature to his rewriting of the chronicles, thus transforming the chronicles into ethnographic description and the missionary grammars into linguistic inquiry. Such writing linked the literary and the linguistic with the ethnographic. The ideology of language that emerged in this process of reinscription is twofold: first, one that upholds, in the celebration of feminized abundance, an originary ontology of America that cannot be ascertained either through revelation or through science, but only through a literary aesthesis of excess; second, an ideology of language that finds the limits to such celebratory aesthesis in the problems raised for native American Spanish speakers with the admission that indigenous languages are contemporary viable, valuable tongues. This latter question is the one that ultimately frames Uricoechea's new scholarly quest, the one that guides his edition of indigenous grammars.

During his stay in Colombia Uricoechea had traveled to the department of Meta in Eastern Colombia, to learn indigenous languages as a means to understand the origins of America. And what he encountered is the way indigenous languages, as a means to understanding origins, raise the conundrum of belonging of Hispanic American Creoles:

> During my trip to Meta, with written texts in hand, it was impossible to make myself understood by the indigenous tribes I visited. So I used interpreters to learn how to pronounce well, and I would make them repeat a phrase as many times as I thought I needed in order for *me to believe* that I was pronouncing it well. If a moment later I needed to use it, it was impossible to make myself understood, even by the same interpreter, when repeating the phrase to him. There are pronunciations that are so different and differences in sound that are so delicate that only a long practice, in the absence of orthologic analysis, make it possible for *a foreigner* to learn an American tongue. (Uricoechea 1871, xlix, emphasis mine)

The geopolitical relation between nation and language characteristic of German philology was not possible for Uricoechea. A native by birth in the Americas but a foreigner by language, Uricoechea names the crux of the matter: an ontology split between place and language. The place of birth is simultaneously an outside, generating a metaphysics of dubious presence. However, since the problem of origins is manifestly unsolvable it potentially gives rise to an eventual linkage to mythology since "it is from the impossibility of beginning with one's own beginning that one can explain the origin of the mythical activity that appears inseparably from the phenomenon of culture" (Sloterdijk [1988] 2006, 41). A metaphysics of dubious presence gives rise to the myth of a portentous continent in terms of species abundance and topographical riches rendered so by literary cultivation. Its limit is the impasse of the relation between hearing a language, pronouncing it, and inscribing it. If the number of modes of articulation of sounds found in different languages of the world seems unlimited (as seen in the first page of this chapter), as are also the entities of nature, the possibility of articulating such sounds properly in pronunciation is the limit. What is revealed by such a limit is the "double character, simultaneously acoustic and mechanical, that makes of the act of language a phonic fact" (Maniglier 2006, 115). Thus, "in the study of isolated sounds . . . the acoustic quality of the phoneme is not a problem; it is fixed by the ear; in terms of articulation one has all the liberty to produce at one's discretion.[10] But once the problem is to pronounce two combined sounds, the question is less simple. One has to take into account the possible discordance between the searched effect and the produced effect; it is not always in our power to pronounce what we want. The liberty of bringing together phonological species is limited by the possibility of tying together articulatory movements" (115).

So it is in the disjunction between the ear and the mouth, between audition and articulation, between assembling sounds or using them singly, between aurality and orality, that the particularities and limits of difference emerge. Here, in such acoustic disjuncture is one of the keys to the relation between the literary, the musicological, and the ethnographic as intertwined disciplines in Latin America and the Caribbean. Let us look more closely at the modes of narration that give us the keys for the significance of such a disjuncture in both Uricoechea's own introduction to the Amerindian grammars and in the missionary Amerindian grammars themselves.

All the twenty-five American missionary grammars reedited in the *Collection Linguistique Américaine* begin, after their new introductions, with the problem of the disjuncture between the Spanish alphabet of the missionaries and the indigenous sounds that could not be inscribed by such an alphabet. Let us take some examples to explore the narratological resources that fill in the void left by the distance between auditory perception, the limits of "tying together the articulatory movements" of the mouth in pronunciation, the ear, and linguistic inscription.

After the lengthy introduction by Uricoechea, the historical grammar of volume I of the *Collection Linguistique Américaine*, the *Gramática Chibcha*, opens with a one-paragraph chapter "On Orthography," in which it is stated that since indigenous peoples had no writing, the missionaries will use "our letters" to write the language, "except Indians do not use D and L, and the R is only used in certain words, and then it is not pronounced harshly but smoothly" (Uricoechea, ed. 1871, 1). The second chapter is "On Pronunciation." The author of the grammar has six rules of pronunciation. Here are the first three: "The first is the pronunciation of the z, which is done by taking the tongue closely and suddenly (*arrimando la lengua de golpe*) and pronouncing with strength. The second pronunciation is that of these syllables, cha, che, chi, cho, chu which cannot be pronounced with the full tongue but with the tip of the tongue only. The third pronunciation is one that is neither that of E nor that of I but in between both, which we write with the Y" (Uricoechea, ed. 1871, 1).

First, the problem of mismatch between alphabet and linguistic sounds is addressed by describing which letters are lacking in indigenous languages and which in Spanish, locating the difference in the disjuncture between acoustics and the sign. Then the author attempts to solve the acoustic-articulatory-interpretive problem of ear-pronunciation-inscription through an explanation of the articulatory physiological mechanisms to pronounce the different sounds. So the phoneme is not only *described*, its articulation needs to be *explained* (Maniglier 2006, 114). The disjuncture between the sign and sound does not stand as an isolated element but as one that is accompanied by a narration of the explanation of how to achieve linguistic competence.

Another one of the grammars *Arte y vocabulario de la lengua Chiquita* (re-edited and published after Uricoechea's death as volume 6 of the collection, by then renamed *Bibliothèque Linguistique Américaine*), uses a longer narrative device to describe the same linguistic disjuncture:

> Not only do the Chiquitos pronounce their tongue well and clearly, but they also do it softly (*con suavidad*), in such a way that there are few in America that have a less difficult pronunciation. This can be observed by anyone who tries to pronounce the words herein written, albeit with a warning that all the letters have in this tongue the same strength and pronunciation as in Spanish except two. The first is z that is pronounced in Chiquito as if it were written ts. The second is x that is pronounced like two ss, with one of them strongly aspirated, in the way Germans and French pronounce sch (I). Spaniards lack this pronunciation and so when they began writing in Chiquito they found no other letter than x to express it. (Adam and Henry, eds. 1880, 2)

Many other resources were used in the indigenous missionary grammars to account for such a disjunction. Smith-Stark, for example, finds "seven aspects of phonological description" for describing such a disjunction in grammars of New Spain,[11] "especially those produced during the first century of grammar writing (1547–1645). . . the graphic representation of sounds, the articulatory description of sounds, the concept of a sound system, the relation between graphemes and phonemes, phonotactics, argumentation, and phonological processes" (2005, 12). Evidently, what has been dismissed historically as simply the "descriptive linguistics" of missionary grammars entails the development of a conceptual framework (Smith-Stark 2005) on the relation between translation, linguistic sounds, their pronunciation by natives and foreigners and their inscription. Such so-called descriptive linguistics generated by the production of new grammars emerged through colonial expansionism in Africa, Asia, and America beginning in the fifteenth century and accounted for a type of linguistic conceptualization between the fifteenth and the nineteenth century that has generally not been taken into account in the general history of linguistics (Pineda Camacho 2000; Smith-Stark 2005), a history described almost exclusively from the German-Asian connection or the simultaneous developments in France. For William F. Hanks, the production of these grammars effectively meant the rise of new types of genres generated by the transformations produced by the missionary enterprise (2010).

I will not dwell here on the particularities of such missionary grammars, and whether their model of production comes from Nebrija's Latin or Spanish grammar or from other models.[12] I am more interested in the way a particular history of comparativism and of difference, articulated through the relation between aurality (what is heard or references the ear) and orality (what is pronounced or references the mouth), as it emerges in this history. By the nineteenth century, when Uricoechea utilizes these grammars to ask questions about the nature of the American continent, he has with him not only the inheritance of the comparative projects of Hervás y Panduro and German comparativism but also a rich linguistic tradition of thinking difference as a problem located between aurality, orality, and alphabetic inscription. The problem of the linguistic sign is one that is simultaneously addressed as acoustic, mechanic-articulatory (or physiological), and in a narratological mode that intertwines description and explanation. Rather than located exclusively in the disjuncture between sign and phoné, the problem of difference is located in the relation between spoken and written language, between the ear, the mechanics of articulation and the actual sound of pronunciation, and between the descriptive narration of a sound and the inscription of a sign. Thus such a difference exists not only between the signifier and the signified but also between the sign and the narrated theoretical explanation, that is, following Lacoue-Labarthe and Nancy, in the Romantic emergence of the literary as theory (1988).[13]

In the final grammar he edited, that of the Goajiro language written by his contemporary Father Rafael Celedón, Uricoechea reutilizes these conceptual-descriptive resources in his introduction for purposes of comparison between American tongues. In a section of the introduction entitled Comparative Study of the Goajiro (*Estudio comparativo del Goajiro*), he compares "the books that deal with the tongues recognized as Carib: caribe, galibí (calibí, caribo, or guaribe), and cumanagoto with its dialects chaima and piritu; besides one of the tongues known with the name of aruaca; and finally, the achagua, guaraní and tupí in order to inquire the family to which the language I publish today

belongs" (1878, 26). As in the previous examples, he presents the comparative results of the relation between alphabet and sounds in terms of a lack:
"all these tongues lack the following Castilian sounds: *c* (before *e* or *i*), *f*, *ll*, *x*,
z" (26), noting also which sounds are missing in the specific languages he is
comparing. Then he does the opposite: he explores which sounds are present
in the indigenous languages but not in Spanish: "Except for the tupí, guaraní
and achagua, all the other tongues have the following letters that are missing in
the Castilian alphabet: *ö*, *ü*, *sh*, equivalent to the French *eu*, *u*, *ch*, and a special
r characteristic of Carib tongues" (27). In this case, the device for explaining
a different sound is that of creating a particular sign for it and then comparing
its sound to a known European foreign tongue. This is the same device used
when describing sounds that are unknown because they have been lost. The
problem of sign-phoné in relation to articulatory physiology-narration is thus
extended into the way the archive reveals American history: "We know from
Father Breton that the Carib language had the guttural *g* and *j* of the Arabs, but
we ignore how many more sounds these tongues had, because the early (antiguos) writers did not write them down, which does not mean that they did
not exist: it is just that the study of phonetics was not yet as profound. We only
need to cite what Father Breton himself says: 'There are savages that speak
so strongly between their teeth and so nasally, that one has a lot of trouble
understanding them' (cited in French)" (27).

So the disjuncture of hearing, pronunciation, inscription, and narration
is not only a question of the contemporaneous significance of the linguistic
archive but also of reinterpretation of its historical significance. If comparative
philology is understood as the field through which history replaces revelation
as the means to transform the understanding of cultural difference, here the
limits of the understanding of culture as history are revealed by the historical
significance of the impossibility of language having a clearly defined scientific
object of study. We "need only cite," as Uricoechea said, a description of the
disjuncture between phoné, articulation, hearing, and meaning to get the full
import of the impossibility of using that as a means to understand what those
indigenous languages were and therefore to use them to define the culture and
nature of the American continent.

Uricoechea is a person who always seems to be located on the brink of a
new formulation. Deeply steeped in comparative philology, natural history,
and antiquarianism as his formative fields, he pushes the boundaries of his
perceptions to the limit, yet his questions remain bound to those for which
the comparative methodology was designed. This happens because he not
only inherits different epistemic resources but because he constantly contrasts

these against the evidence given to him by his own experiences. This comes up again and again.

In the rest of the study, for example, and in the spirit of Hervás y Panduro, he compares different aspects of the languages named above: numbers, plural endings, formation of the degrees of comparison, gender, pronouns, conjugations (which he transcribes not only in Latin alphabet but also in Arabic script!). He does all this in order to be able to organize the languages into "language families." However, even though in classical comparative style, Uricoechea makes lists of the same word across different languages, he is acutely aware of the problem of meaning that arises in translation: "I know by experience that in the majority of the vocabularies a radical word alone does not exist. Rather, it is found with a suffix or prefix that forcibly changes the syntactic relation between words. . . . It is most difficult, almost impossible to make an Indian translate an isolated word or its absolute signification" (Uricoechea 1878, 41). He states these problems with translation in more detail in the introduction to the Chibcha Grammar,

> We easily know that if our lexicographers believed they were writing a Chibcha dictionary, they only did a dictionary of translations or explanations, a vice we still find among modern authors who are happy with explaining a term without reproducing the equivalent in another tongue. The great efforts they make in a forced translation that does not carry the imprint of the Castilian language are well known . . . and the obstinacy in translating devil, soul, and such abstract things that they themselves could not make the Chibchas understand, who, without doubt, would form an entirely different notion of signification to the one that the conquerors wanted to give it. Soul, for example, is translated as *fihizca*. Since the Chibchas did not know the entity that the Castilians call soul, they did not have the word to name it. And in naming it *fihizca* which means air (*huelgo, aliento*), they without doubt understood the subtle part, the air that is breathed, but they could hardly understand that for such a thing they had to bear the many sacrifices imposed by their new lords. (1871, xlix–1)

He thus comes to pose the problem of language as a clearly definable entity, creates a linguistic relativism, an Americanist version of the Sapir-Whorf hypothesis, and the implications of that for the interpretation of the colonial experience.[14] Uricoechea begins to develop a theory of language that is articulated to personal experience, as an empirical analysis of the problem of letters and sounds. This is developed through a series of analytical and narratological devices for describing the comparative problems generated by the disjuncture between heard and written languages in order to generate a theory of linguistic relativism, but one tied to its significance for theorizing Latin American difference. What Uricoechea discovers is that there is a difference between the abstract, historical comparison of languages used to create "language families" to understand the nature of the American continent and its peoples, and the reality of the actualization of language itself, both in the acoustic-mechanic sense of hearing-pronunciation and in the syntactic-semantic of production of meaning. At this point, one could apply the words Maniglier uses for Saussure when describing what he discovers when phonologically analyzing the procedures of speech: ". . . there are realities, that are such that they can only be understood as actualizations (which in the end, means pronunciation) and that therefore demand the separation between virtual entities and processes of actualization. There are parts of the real that are only real by the fact that they are actualized" (Maniglier 2006, 125–26).

But Uricoechea does not seek to theorize the implications of the perception of this distinction between the actualization of language and its abstraction as an entity in order to develop a theory of language, but rather in the service of a theory of America's ontology. Let us recall that this particular relation between the virtual and the actualized in language is also related to the disjuncture produced by being born to a place of origin but to a "foreign" mother tongue. Thus, in Uricoechea's theorization, such linguistic inquiry is not meant only (and not meant primarily) to understand the nature of the linguistic object, but he uncovers the nature of the linguistic object in the service of trying to uncover the nature of the American continent for producing an Americanist theory. Thus, his theoretical purpose is revealed by understanding how he embeds the linguistic problem posed by indigenous languages of articulation of sound, articulation, and sign in a series of narratological devices that link the exploration of language to that of the nature of America.

As stated earlier, the three grammars Uricoechea edited are prefaced by his own account of the indigenous group to which the grammar belongs. Such an account is composed by rewriting the materials he has read from the chronicles as a narration in third person. One has the impression of reading an account written by a person who was witness to the historical events narrated. He uses the third person of the ethnographic account and simultaneously, by the use of the third person, upends the dialogic implications of the use of the first and second person (Esposito 2008), relegating indigenous persons to a distant other, as if *they* were the remote inhabitants of America, not only in time but in place. Thus emerges the complex relation to anthropological discourse in the public sphere in Latin America, as that which simultaneously exposes and upholds the particularities of indigenous expression yet which, in its relation to the problematic metaphysics of presence of Creoles, torn between being born in the region yet of foreign tongue, finds its limit in the politics of recognition of indigenous peoples and affairs in the constitution of the public sphere.

In Uricoechea's introductions to the three grammars he edits we learn about the geographic localization of the particular indigenous groups to which the grammar relates, their state at the moment of conquest, myths, rituals, uses and customs, and finally, his own appraisal and theorization of the language and grammar he edits in that particular volume. As we saw, in the first volume all this is prefaced by the paradisiacal celebration of American abundance of materials, species, and climates, which sets the purpose of the whole collection, in which grammars are to appear as "monuments" that speak to such a rich history of speciation and topographic transformation. So the ethnographic third person for describing indigenous cultures and peoples is entwined with a maternally encompassing and welcoming literary description of a portentous nature that receives the colonizer. While the original inhabitants are othered and circum-scribed into the third person, American nature receives the colonizer as her own.

Let us recall that in the case of Uricoechea (or of America in general), the problem of place and language for the Creole subject cannot be solved in the proposition of the generation of an identity. Thus, the double disjuncture posed by the difference between the virtual and the actual in language, and the difference between place of birth and mother tongue, is resolved in the literary relation that emerges as mediating both and in the trope of ethno-graphic distance and othering that circumscribes indigenous peoples. The relation between nature and culture that emerges in such a mediation is that of a narrative reason in which the magnificent descriptions of nature are in-tertwined with the ethnographic exploration of the local in the creation of a literary language that stands in for the difference in the disjunction between aurality and orality, made evident by the aural components of language. What emerges here is an ontology of hyperabundance and constraints, one in which the diversity of species that characterizes tropical America, the part of the world with most species diversity, constitutes a hyperbolic real whereas the disjuncture between place, speech, the ear, and writing constitutes its limit. This makes the relation between the irreducibility of the given in language and the irreducibility of the given in speculative reason, manifest.

It is in the relation between the conscientization of problematic origins, histories of life, and histories of language that America emerges as a continent in which the relation between what we call mythology, nature, and language structure its ontology. If one direction of such an ontology is its eventual hyper-celebration in magical realism, another is its potential use in essaying "materials for a future history of imagination and affectivity" (Derrida citing Richard [1978] 2005, 2), and yet another is its role in developing a particular history of comparativism in the region.[15] Such a project is located between the aesthesis of the literary, the theorization of the mythological and anthropological and the impossibility of reducing the acoustic to a single domain due to its constant movement between the virtual and the actual.

One wonders where further theorization would have led Uricoechea. His premature death and the loss of a great number of his writings have made him, until recently, a relatively obscure figure in contrast to the foundational political and epistemological role of other nineteenth-century Latin American philologists. But perhaps his relative neglect is due to Uricoechea's caution in debating questions regarding the political theology of the lettered city. Jorge Isaacs instead used his ethnographic ear to overtly challenge the confessional acoustics underlying the lettered city's political theology.

The Confessional State and the Ethnographic Ear: Isaacs's Challenge to the Politics of Conversion

When Isaacs published his *Estudio sobre las tribus indígenas del Magdalena* (1884), he had already overtly challenged the conservative political establishment in numerous ways. After having lived for several years in Bogotá, participating as a member of congress for the Conservative party, in 1877 he returned to his native state of Cauca. He changed political allegiances, joined the radical liberals and began working as secretary of government for the Cauca state. He promoted an explicit anticlericalism, the separation of church and state in all governmental affairs, and the establishment of a secular public education. In congress he vehemently opposed the assignation of rents to priests. In 1876 he fought alongside his cousin, philologist César Conto, who was president of the Cauca state from 1875 to 1877, in the battles between radical liberals and conservatives.[16] In 1880 Isaacs led a radical insurrection in Antioquia against the conservatives, only to be defeated a few months later. As a writer, he promoted the cause of radical liberals in his *La revolución radical en Antioquia* (1880) (a political explanation of the causes of the uprising), in militant poems, and in his work as a journalist. If Uricoechea's passion took the form of an obsession with all things American, Isaacs's passion took the form of an overt political and military contestation against the instauration of a clerical state by the conservatives.

In 1881, President Rafael Núñez, who was then a moderate liberal, named Isaacs secretary of the scientific commission that was to finish the work initiated by the Agustín Codazzi Geographical Expedition (1850–1859) by studying the natural resources and geography of the departments that had been left unexplored, beginning with the Caribbean state of Magdalena. The scientific commission did not prosper due to internal problems and lack of financial support, but Isaacs finished the labor he began with "no other stimulus but the goodness of the aboriginal peoples of those regions and the desire to honorably fulfill the duties assigned to me" ([1884] 1959, 15). His *Estudio sobre las tribus indígenas del Magdalena* was the result of this research. This is not his most militant or politically vociferous writing, but it is, perhaps, one of his most radical works. Besides denouncing governmental abuse of indigenous groups in the Magdalena region,[17] Isaacs proposed the secularization of indigenous affairs within the nation-state by transferring their administration from the hands of missionaries to that of ethnologists and archaeologists. This implied rethinking epistemic assumptions about indigenous languages, myths and history, and the juridical status of the personhoood of indigenous people. Here the question of Darwinism is once again crucial. In this case, it is mediated by Isaacs's conversations and travels among the indigenous groups of the Magdalena region. He uses it to compare the information they give him with his own understandings of language, history, mythology, and the natural sciences.

Darwinism provided the possibility of moving between ideas about language, antiquities, and the natural sciences. Such translation between fields was what made possible the anthropologization of the political (Latour 1993) by moving the discourse on indigenous groups from one of reduction (*reducción*) and Christian policing (*policía cristiana*) to one of ethnographic science. In Isaacs, the question about (biologically defined) life was enmeshed in the question of (anthropologically and linguistically defined) indigenous peoples in the political search for the particularities of the American continent. His inquiry was initiated methodologically by his auditory willingness to validate and acknowledge indigenous knowledge acquired in conversation with indigenous persons. This led him to question the authority of priestly chronicles and grammars. Such privileging of knowledge by indigenous persons and ethnographers over that of priests provoked Miguel Antonio Caro's irate response: "[Isaacs] proposes that the savage tribes be domesticated by archaeologists, ethnographers, philanthropists and missionaries, all together and the missionaries in the last place. And not just any missionaries but only those that Mr. Isaacs approves. Praise, blame, doubt, mockery, absolute lack of hope, all of it is mixed together there, and in the end Mr. Isaacs does not propose anything, and does not even know what to propose to the government. This is the sad and deplorable consequence of lost faith!" (Caro [1886] 1980, 1087).

Caro's response is more reminiscent of accusatory inquisitorial judiciary procedures regarding lack of faith than of an intellectual disagreement. *Estudio sobre las tribus indígenas del Magdalena* was written and published during the rise to power of the Conservative party and in the years immediately prior to the period known as the Regeneration (1886–1899) which implemented the terms of national unity proposed by the conservatives. Catholicism and language became crucial cornerstones of national unity for them (González González 1997; Von der Walde 1997; Sierra Mejía 2002; Arango 2002; Rodríguez-García 2010). Their political ideals were inscribed into the constitution and into law during the late 1880s and 1890s. They were to have sway during the following thirty years of rule of the conservative party and to determine, in good measure, the relation between the political and the juridical in Colombia for most of the twentieth century.

During the Regeneration and its aftermath Colombia had a series of philologist (or grammarian) presidents, and the discussion of grammatical issues such as the proper use of a gerund often occupied hours of congressional debate (Deas [1992] 2006; Rodríguez-García 2010).[18] But as Malcolm Deas warns us, one should not collapse the link between academic philological interests and public power into Benedict Anderson's explanation of the rise of a vernacular language as a crucial foundational element of nationalism ([1992] 2006). The phenomenon we are facing here, in terms of the discussion of indigenous languages, is that of the use of a profound knowledge of a discipline, philology, in forging the political definition of the indigenous person for the rule of law. The most influential political figure in this process was Miguel Antonio Caro.

Miguel Antonio Caro (1843–1909) was born in Bogotá, a city he rarely left, not even to visit other places within Colombia. He was the son of the founder of the Conservative Party, José Eusebio Caro. His education proceeded through formal schools in Bogotá and through private tutoring and included philology, Latin, and training in a strong Hispanist tradition based on the relationship between language and religion. From August 1871 to August 1876 he directed the newspaper *El Tradicionista* (The Traditionist) where he consolidated his ideas against radical liberalism, then in power, and where he began to conceive of a political party based on Catholicism. During the 1880s he shaped, along with President Rafael Núñez, the major political tenets of the Regeneration and spearheaded the constitutional process that led to the Constitution of 1886, of which he is the main author and intellectual figure. Elected vice president in 1892, Caro held power until 1898. Though he never formally assumed the presidency and maintained the title of executive vice-president, Caro exercised the powers of that office following President Rafael Núñez's illness and subsequent death in 1884. He wrote prolifically on philosophy, religion, philology, and politics. As stated by Von der Walde, "in his literary, philological and political writings, he elaborated a coherent conservative thought with solid bases in history, religion and language. Over such an edifice he managed to create his version of the anti-utilitarianism that was then sweeping Latin America, converting it in a coherent anti-modernist discourse" (1997, 72).

During the months preceding and following the final draft of the Constitution,[19] Caro was particularly concerned with the relation between religion, language, and nation. The new Constitution made Catholicism the only official religion of the nation and gave the Catholic Church the right "to administer its own business and execute acts of spiritual authority and ecclesiastical jurisdiction, without the need of authorization from civil power" (article 35 of the 1886 Constitution). A treaty granting the Vatican and the Catholic Church extensive political powers was signed in 1877, and public education was officially turned over to the Catholic Church. In the months following the drafting of the constitution Miguel Antonio Caro published, in the Diario Oficial (the official newspaper of the government), El Darwinismo y las misiones (Darwinism and the Missions), his bitter, virulent response to Jorge Isaacs's Estudio sobre las tribus indígenas del Magdalena, which had also been submitted to the Diario Oficial in 1884 but only circulated in 1886.

Caro's Darwinismo y las misiones is a long diatribe (more than fifty printed pages!) against Isaac's Estudio sobre las tribus indígenas del Magdalena. His outright hatred of this text is centered mostly on Isaacs's secular humanism and it allows us to see how the religious critique generated by the relation between the study of indigenous languages and questions regarding the origin of America was constitutive of the politico-theological. It has been remarked that Caro is responsible for the creation of a "confessional state" (Arango 2002, 125). Such an idea rests not only on the multiple ways in which religious power was inscribed into the Constitution of 1886 and into treaties and laws (which would make it a Catholic state) but also in the way in which the idea that deviation from religion as public authority was sinful was deployed through a carefully constituted network of scientific, political, grammatical, and pedagogical discourses and practices.[20] The implementation of such an idea had to do with the astute political use of language and the rhetorical style Caro employed in the art of polemic argumentation, so crucial for the deployment of governmentality. Such knowledge of linguistics and rhetoric was skillfully used in responding, point by minute point, to Isaacs's secular humanism. The polemic was centered, in good measure, on a contrasting politics of listening to indigenous languages and peoples. If the authoritative-sensory concretion of the carceral state is the panopticon, the authoritative-sensory mode of the confessional state is the intimate perversion of the priestly confessional booth.

The polemic emerged as part of the history of transformation of audition in the redefinition of the relation between the secular, political secularism, and the politico-theological. In his pioneering study, Alain Corbin showed how the (often forceful) transformation of the role of village church bells during the French Revolution entailed the desacralization and resacralization of space and time by altering the way "the culture of the senses" was enmeshed in people's daily routines, in the politics of labor and leisure, and in the definition of the public and the personal (Corbin 1998). The polemic between Caro and Isaacs is another historical instance in which we can see the relation between transformations of the sensorial, juridical regimes and the redefinition of the public sphere. Here at least four auditory regimes or practices were intertwined: The idea of listening as revelation, historically built on the capacity to "hear things" that are beyond scientific or empirical confirmation (Schmidt 2000) crucial to the deployment of faith as a moral and epistemic tenet; that associated to the scientific practice of ethnology that is methodologically based, purportedly, on hearing "others"; the regime of the confessional nation-state in which hearing plays the role of bringing people under disciplinary coercion; and the efforts of indigenous peoples themselves, who necessarily have to accommodate all these auditory practices to their own and deploy their own strategies to restructure their own auditory uses and knowledge in the place assigned to them in the emerging public sphere of the nation. What is at stake ultimately in the auditory configuration of the public sphere is the political definition of the person, both indigenous and nonindigenous. Thus, the question is not only historico-political but also philosophico-juridical (Anidjar 2003) and raises the issue of how the force of the law is used to settle an ontological question (Ludueña 2010) when it is couched between struggles about the definition of life, the sensorial, and the epistemic.

The tension between these different auditory practices and their way of structuring the public sphere was brought to the foreground by the way the audibility of indigenous languages challenged the lettered city's aspirations to grammatical purification. Talal Asad distinguishes between the secular as an epistemic category and secularism as a political doctrine (Asad 2003). If Isaacs proposed a different epistemic relation to indigenous groups, his secular humanism also reveals the constitutive relation between the secular as an epistemic category and the internalization of the sense of revelation present in the privatization of religious feelings that a secular politics entailed. In the intertwining of the revelatory, ethnographic, and confessional auditory modes, what we see are the "markers of a historical shift" (Anidjar 2008, 18) that, while politically manifesting as the opposition between a secular humanism and the confessional state, were conceptually underscored by politico-theological elements characteristic of an enlightened acoustics of rationalization. I do not mean, by proposing such a common ground, to erase the differences between the juridical implications of the acoustic biopolitics that defined Isaacs's secular humanism and Caro's dogmatic Catholicism, especially with regard to the political definition of the indigenous person. But rather to explore the complex intertwining of different auditory regimes in the relation between the political theology of secular humanism and that of the confessional state. This section is about exploring how such auditory history and its philosophico-political implications unfolded through the particularities of the polemic between Isaacs and Caro regarding the acousticity of indigenous languages.

ON HEARING INDIGENOUS LANGUAGES

Isaacs's *Estudio sobre las tribus indígenas del Magdalena* is divided into two large parts, one on indigenous languages and one on geography and history. Although the first part was initially intended as an appendix, it appears as the initial part of the text because it was the most difficult to lay out typographically, so the editors needed to begin its publication process first. This inverted the order and importance assigned to this text by Isaacs since obviously it was not meant as a lengthy grammar but as something to be attached to the central study on geography and history. This intended change of order was already a significant move. Philology here begins to become an attendant category that informs history and anthropology rather than the other way around. As such, the section "on linguistics," as Isaacs called it, is not a detailed study but rather an assortment of multiple indigenous languages of the region, gathered and inscribed as either fragmentary "samples" or somewhat more complete "studies." It is divided into four parts: a study on the Businka (today Ika) language, a sample of the language of the Chimila tribe, a sample of the language of the Motilones Indians, and a study on the Guajiro language, all of them languages of indigenous groups of the northern Colombian Caribbean.

Isaacs's politics of hearing brings together an incipient ethnographic intent and an appreciation for the sound of the different languages. For Isaacs, it is not only necessary to hear an indigenous person to understand the nature of indigenous languages. For him, Spanish speakers are unable to imitate them by lack of articulatory training in such sounds: "The Chimilas' speech is notable for being guttural and nasal, inimitable for those of us who speak a clear and resonant language like Spanish. One needs to hear an indigenous person in order to form a precise idea of those sounds" ([1884] 1959, 59). Such a mode of listening is what gives him a new form of authority vis-à-vis contemporary grammars that follow the model of the missionary ones. Isaacs compares his own findings with the Goajiro Grammar compiled by Father Rafael Celedón, a missionary priest in the immediate years before Isaacs's exploration of this region.

Celedón's grammar had been published in Paris by Ezequiel Uricoechea in 1878 as volume 3 of his *Collection Linguistique Américaine*. Isaacs's critiques are several. First, sonic: In Rafael Celedón's grammar we find the letters o, u, that are meant to represent the French sounds eu and u respectively. But, says Isaacs, "We confess to not having found them and presume they do not exist, because we have heard pronounce u, and rarely o, e, in the words that the author says they figure" ([1884] 1959, 67). Isaacs deauthorizes a priestly grammar based on his acoustic competence. Caro, of course, finds this deauthorization of Celedón's careful study totally unacceptable. In response he weaves his overt critique of Isaacs's dismissal of priestly privilege with his phonological authority on European languages:

> The French mute *e* is the echo of perfectly sonorous syllables of a mother tongue and is characteristic of a derived language that is totally composed of words with accent on the last syllable, and that has no words with accent on the penultimate syllable except those that finish on such an *e*. It is more than doubtful that the businka, a language of strong and precise sounds, full of words (*voces* in Spanish) that have their accent on the antepenultimate and penultimate syllables, possesses the delicate subtlety of the mute *e* that Isaacs imagines having perceived. (Caro [1886] 1980, 1059)

In Caro's response phonetics becomes the means of a highly racialized linguistic argument. Any segment written by Isaacs that hints at delicacy and elegance of sounds is immediately repelled because a barbarous tribe is incapable of such sonorous attributes. One more example among many will suffice. Caro quotes Isaacs: "The particles *kar, kor* are sometimes used only to give more elegance to the expression" (Caro citing Isaacs [1886] 1980, 1059). To which Caro responds: "We cannot conceive how Isaacs, in a few weeks, managed to perceive elegance in the diction of barbaric languages. Not that barbaric languages do not have occasional emphases; but such accidents could hardly have been appreciated by Mister Isaacs, and in no case as literary *elegances*" (emphasis in the original) (Caro [1886] 1980, 1059–60).

Such a move requires a translation from phonology to the literary as that which holds the key to linguistic authority. Caro has to admit that indigenous languages are logical phonological-semantic structures since that is precisely the ground for a global comparative linguistics and for the production of missionary grammars. Indigenous peoples cannot be dismissed as nonhumans because, like other humans, they speak a language. But such languages do not possess the refinement that makes literary expression possible. So it is upon the critique of acoustic elegance and literary expression that Caro bases his conception of "barbaric" peoples. This contrast between modes of valorization of the sounds of indigenous languages between both scholars is also constituted through other linguistic aspects.

Isaacs has a section that outlines phrases "captured while on board" that denote the corrupting influence of Antillean traders on indigenous languages of the Caribbean that at that time were increasingly being mixed with Papiamento and Spanish. While he listens for their elegance in diction and literary possibilities he also laments the current linguistic transformations. Isaacs thereby installs a crucial modernist humanist trope: the secular valorization of local culture while lamenting its disappearance. The documentation of local culture hereby becomes the reenactment of a "repeated ending" (Anidjar 2002). Here the critique of the conquest and of the disappearance of indigenous peoples and their expressions (see chapter 2) is refracted through the humanist concern with the disappearance of traditional cultures through mixture. This romantic humanism is the key to Isaacs's political move: in a secular state, the governmentality of indigenous groups requires the pastoral care of the state as the ruling agent to which missionaries are also subject. Such a move recasts, into the secular authority of the sciences, the concern with purification of subaltern expressive culture that is at the heart of religious authority in comparative linguistics.

Confronted with such a transformation, Caro invokes the politics of religious purification. He blames the degeneration of indigenous languages on the "Antillean Jews" who are corrupting the indigenous groups for both economic and religious motives, tacitly implicating Isaacs's father, who was an Antillean Jew, and raising the specter of Isaacs as a convert: "It would have been good if the harm caused by these ruthless smugglers [Dutch Jews from the island of Curaçao] were reduced only to the corruption of the native language, the only harm deplored by Mr. Isaacs" (Caro [1886] 1980, 1085). We can link this diatribe against Isaacs' position as a convert to Caro's only valid motivation for the study of indigenous languages:

> The missionary learns the barbaric languages in order to catechize the savage; the philologist studies the data given to him by the missionary to discover similarities, inquire about origins and prove linguistic laws. None of them has the motivation, that Mister Isaacs claims to have, of a general interest in conserving *the purity of a language* [emphasis in the original] that has never reached it, nor carries within it traces of ever acquiring the refinements of the literary and classical languages— bare trunk that does not bear flowers. The commitment that Mister Isaacs has shown on taking down philological notes on the Goajiro language . . . is an exceptional case in the history of this genre of research. ([1886] 1980, 1090–91)

Isaacs's history as descendant of a convert reemerges this time, years after the concatenation of silences that marked the interpretations of *María* (see chapter 2), tacitly invoked through the general background reference to Antillean Jews in the struggle between religious and scientific authority. Isaacs is aware of the pioneering nature of this book meant for "ethnographers and archaeologists . . . if it indeed merits such a honor" ([1884] 1959, 15). His secular humanism challenges the political theology of missionary linguistics and repositions him in that of the nation-state by invoking ethnographic and archaeological sciences. Moreover, this is precisely the brunt of his critique of the contemporary indigenous grammar on the Goajiro language produced by Father Celedón. It "does not inspire any trust in us as much because it was not made in the midst of the tribes that speak this language as because, during our stay there, we had occasion of noticing many mistakes consigned in this work" ([1884] 1959, 74).

To extend such scientific criticism further, Isaacs ends his linguistic study with a fragment in which he transcribes very different translations of Our Father in order to demonstrate their questionable authority as translations. In doing so, he again deauthorizes both the philological lineage and the missionary one based on the comparison between written collections of such translations. The move from a comparison between translations of Our Father, characteristic of eighteenth-century language studies, to careful attention of the process of translation itself, positions him more as a linguist than a philologist. This move from comparative grammar to linguistics is related to the emergence of a Romantic anthropology based on Isaacs's interaction with indigenous groups:

Detailed research, obstinate effort: first, to capture the affection of chiefs and priests and their families; then to travel through the terrains that they inhabit and the deserts where they reign, in their company; in the study of languages, not to lose one instant that leads to the acquisition of valuable data, a new word, a particular linguistic turn; obtain from the elders, through gifts, benevolence and astute patience, that which has not been easy to obtain from the chiefs and medicine-priests regarding traditions and religious beliefs; conquering the affect of women, usually distant and distrusting, through gifts of beads and trinkets that they greatly estimate to adorn themselves, caressing the children, appreciating the women elders. Days and nights, losing the count of the number of days and dates, with no other society than that of barbaric peoples, with no other roof or home or care than theirs. As horizon, that which has not been seen, the grandiose, the ignored. ([1884] 1959, 24)

The indigenous person is no longer one to be converted but an informant, an aide to the constitution of a literary and scientific paradigm instantiated by listening to the aesthetic and scientific elements of language and to the data the informant provides. It is easy, of course, to deconstruct the previous passage in terms of denouncing this exchange as one of a highly unequal paternalistic politics of benevolent power where Isaacs intones an indefatigable and permanent search for any detail that might provide "valuable data" through the instrumental deployment of "gifts, benevolence and astute patience" to each and every indigenous person who might provide information, all against the backdrop of a yet to be discovered grandiose nature and geography. But to limit our critique to such things would be to ignore the contrastive attitude and mode of writing between Isaacs and his Creole contemporaries, who primarily privileged their disgust at living Indians by either denying their contemporary existence or affirming the need to convert them to make them viable persons. Isaacs instead questions the grammars, travel books, geographies, and histories of the period through the information given to him by indigenous persons. Historical texts function less as chronicles to be revealed and cited and more as sources to be questioned by ethnographic experience. This is not only a change of paradigm of research on indigenous languages and peoples. Isaacs proposes a move from the truth of religion to the truth of science (Latour 1993) as that which constitutes the significance of the indigenous person in the secularization of the modern, a dispute between different notions of truth over the rule of indigenous peoples that continues to the present.

To sum up, the types of translation enacted in this polemic are between the acoustic dimensions of indigenous languages and their inscription, between the phonological and the literary, between the political theology of missionary grammars and the political theology of an emergent secular Romanticism, between the significance of indigenous language and the significance of indigenous peoples, all in the service of abstraction of different theories regarding indigenous peoples in order to define the proper mode of authority over them in the nation-state. This process of "abstraction of the pragmatic life of social texts is a critical moment in which national ideologies are localized" (Povinelli 2002, 188). Here we find such a politics of abstraction is crucial to defining the relation between the shared but incommensurable politics of life between indigenous peoples and nonindigenous ones. But such a process of abstraction, in turn, needs to be transduced into a politics of life construed not only through linguistic dimensions but in the relation between linguistic dimensions, biological heritage, and territory. Thus the discussion moves from the materials of language to those of history, evolutionism and geography. The different politics of life that emerge from this abstraction calls for a "transductive [historical] ethnography," that is "a mode of attention that asks how definitions of subjects, objects and field emerge in material relations that cannot be modeled in advance" (Helmreich 2007, 632). Here we can call for a "transductive ear" that helps us "listen for how subjects, objects, and presences— at various scales—are made" (632). We can explore this in the contrast between Darwinism and creationism, between biology and history, and between faith and science, as the eminent domains through which the transductions between competing auditory regimes take place. It is through such transductions that the "liberal cycle" (Rivera Cusicanqui 2010, 40) of nineteenth-century governmental politics about indigenous groups gets inscribed into the political theologies of the nation-state and the secular humanism of the new disciplines.

DARWINISM, FAITH, AND THE
AUDITORY REGIMES OF TRANSDUCTION

For Isaacs, "the New World is an immense necropolis of nations that perished in the century of Conquest and studying them in the remote solitudes that serve as their tomb is the preferred labor of ethnogenesis in contemporary times" ([1884] 1959, 237). This impulse of recovery of the past that again posits the end of indigenous groups as the original moment of interpretation (Anidjar 2002) is based on an understanding of history that seeks to integrate biological transformation (evolutionism), mythology (myths of creation and migration) and archaeological evidence (pictographs and their multiple interpretations) in the service of defining the nature of indigenous peoples.

Isaacs makes a radical move, which is to understand the transformation of life present in evolution and in mythology as crucial to an understanding of history and therefore to a national rather than colonial missionary politics for indigenous groups. To do so, Isaacs transcribes the myth "on the origins of the human species" told to him by a Businka priest as a key to the proper interpretation of the archaeological evidence left by stones and as evidence of the historical migrations of indigenous groups. He searches for data on the historical elements contained in the myth of origin in a stone shown to him as symbolizing the place where the first human couple appeared on earth. But the stone is covered by moss and has no signs. However, another stone has signs. He interprets the drawing carved on that stone as being half simian and half hominoid. As in the case of the myth, his interest in this pictograph lies in the probability of its referencing a remote historical past that reveals the traces of evolutionism. He interprets the pictographs through a theory of mimetic representation in which the footstep imprinted on the rock resembles the supposed ancestral figure:

> If my susceptible readers can tolerate it, those of us who are partial to Darwinian theory could suppose that figure 12 [the figure drawn on the stone], half simian and with a very strange face, is the representation of the form of the animal, scary as can be seen, in the scale toward perfection. Figure 13, which looks like an Ibis whose head is formed by the sign of the Sun, and whose body has an eclipse with another circle, might be of importance for archaeologists who know the details of Egyptian emblems. (Isaacs [1884] 1959, 245)

Isaacs pairs archaeological evidence as a sign of evolutionary theory with the global comparative impulse that locates the Americas and the Orient on similar interpretive grounds. This idea of history involves the history of hominization and entails accounting for the material traces left by evolution and by mythology. As was common in the period, he transfers the methods used in comparative linguistic analysis to archaeology. Isaacs's method for reading pictographs is the same that Humboldt uses: that of looking for representational resemblance in order to interpret symbolic figures. He supports his own Darwinian archaeological interpretations through Humboldt's writings on pre-Columbian archaeology among the Muisca Indians.[21] In making such an interpretation of archaeological signs on a stone, Isaacs questions the chronicles that had interpreted such a stone as a sign of the passing of the apostles through New Granada. His proposition for validating myths and evolutionary theory entails, once more, the devaluation of priestly signs of revelation. Caro, of course, vehemently disagrees since "that which makes a good historian is that they were religious. Period. This is the symbol of scientific trust" (Caro [1886] 1980, 1092). Faith and science are mutually constitutive in Caro, an issue to which I will return.

At the time Isaacs was writing, the separation between the natural sciences and philology was just beginning to be accomplished and the relation between both spheres provided "an entire world of integrative thinking among intellectuals of the Victorian era, an outlook that bid to unite the natural and the human-cultural spheres" (Alter 1999, xi). In some of its aspects, Darwin's evolutionary theory closely resembled the historicist comparative linguistic impulse understood as "the evolution" of languages across time, a resemblance that Darwin himself used in generating his own theories (Alter 1999). In the case of Darwin, metaphor and analogy were the means of comparing the transformation of languages and that of species (Alter 1999). In the case of Isaacs it was the mimetic impression found on rocks, the relation between antiquarianism and philology, and the willingness to consider mythological narrations as historical ones. Thus, both indigenous myth and evolutionism are positioned against biblical revelation.

In Isaacs, the physical evidence is comparatively studied through the methods devised by philology, but the idea of change is transduced from a linguistic-archaeological one to a biological one. This is the crucial step for bringing the notion of the person from the historical-political realm to the juridico-philosophical. But Isaacs did not have the power to enact the consequence of his transduction in the juridico-political sphere. The juridical move from the historico-political to the philosophico-juridical was instead provided by Caro, who had the power to write and enact the law. Through his rejection of Darwinian theories, Caro builds his own theories of the relation between race, tradition, and the indigenous person.

In the first instance he rejects Darwinian theories due to the problem of accounting for heritage:

We do not admit for any rational being, including Darwinian ones, the miserable lineage (*la miserable alcurnia*), that they, who possess such scarce noble sentiment, attribute to themselves. To stamp his hypothesis the writer asks for the condescension of his *very susceptible* readers [emphasis in the original]. So what! Is rejecting a hypothesis that negates our exalted origins and immortal destiny and reduces us to the sad condition of descendants of one of the most repugnant brutes, an excess of susceptibility? (Caro [1886] 1980, 1061–62)

Instead Caro turns to the relation between climate and beings as constitutive of differences between "races." Following Caldas (see previous chapter) he stresses that phenotypical difference is a cause of climate influence and not of heredity, and therefore racial differences cannot appear as examples of a critique of monogenetic religious origins:

In the first place, man in its *animal* part, even considered only as an animal, exhibits the unity of its species, and does not permit its confusion with other animals. Anatomically, there is no difference between the black African and the white European. Skin color and other accidental peculiarities depend on the powerful influence that a long series of ages exercise over the physical organization of climates and other material conditions. The hybrids of plants or animals do not possess the value to reproduce themselves, while the crossings between different races of men are fecund. This demonstrates that we are all one family that received as its heritage all the regions: *grow and multiply and fill the earth.* (Caro [1886] 1980, 1066)

The intense history of mixture between peoples in New Granada does not allow positing racial difference based on phenotype. As animals, as a species, all humans are equal due to the biblical doctrine of Adamic origins. But the notion of blood purity from the colonial era persists in the idea of noble descent, the primary element of the definition of tradition for Caro. Tradition then becomes the biopolitical means of guaranteeing the temporal relation between the past and the present that characterizes all heritage, a move from the biological to the cultural:

Darwinian theory is not a tradition. The "scary animal," supposed progenitor of man, could not transmit its remembrance to his immediate descendant. If among the people that have history, genealogies and traditions, there is no simian tradition, how are we to look for it among the tribes that only conserve confused and fantastic remembrances of their origins? What is more, that tradition does not and cannot exist. Tradition requires two fundamental conditions: the identity of the species across successive generations and its intellectuality. Tradition is the memory of the human race. . . . Tradition, like the language that transmits it, presupposes a permanent *social* state. [emphasis in the original] (Caro [1886] 1980, 1069–70)

So, the difference between animals and humans is that humans have a social life that animals do not have, something that can be accounted for through tradition. Even though all are "one human family" because the different races can biologically mix, not all mixtures are the same and not all races have the same nobility or tradition.

All of the above discussion is finally brought to bear on the politics of the question of incorporating indigenous persons into the nation. For Isaacs, evolutionism as scientific doctrine cannot explain one of the main political problems of social disciplines—that of addressing the difference in power between different "races," a reference, probably to evolutionary doctrines of biological selection of the fittest:

> The anthropologists and sociologists who carry out diverse classifica-
> tion of races explain, in their own way, the inevitable victory of some
> over others, and the extirpation or absorption of the defeated races.
> That might be very scientific. But the history that such affirmations
> could justify demonstrates, at the most, that humanity is very distant,
> almost as much in the last centuries as today, of the perfection or selec-
> tion that it will reach one day, possibly a remote one. Meanwhile, despite
> the redemptory doctrine of Christ, human fraternity, the synthesis of all
> progress on earth, is a utopia. (Isaacs [1884] 1959, 195–96)

Experienced in the arts of war, Isaacs does not give credit either to scientific or religious justifications as a means to overcome the difference in power
between "races." In rejecting the implication of both a religious and scientific
understanding of societal relations, he reveals his capacity for uncovering the ultimate similarity between a scientific justification of conquest and a theological
one. In his secular humanism, he brings the theory of evolution or the "striving
toward perfection" away from biology and into the domain of the social contract.
With this interpretive gesture Isaacs turns individual religiosity into an inward
affair, signaling a shift "from civic man to inward man, from objectivist to sub-
jectivist philosophy" (Casanova 1994, 50) characteristic of secular humanism.

He also brings up the question of "fraternity" as the objective of a secular politics. The impossibility of fraternity anticipates the twentieth-century
characterization of Colombia as a country of "fratricidal wars" (Palacios 1995)
and of violence as its foundational myth. The future appears as permanently
evanescent. Such a building of futurity into an impossible history of contemporary violence presents the problem of the nation and the region as one in
permanent formation and failure. It also constitutes the nation as one that is
perpetually built on the maintenance of an idea of the enemy.

In the rhetorical deployment of the polemic generated by Isaacs's text, the
political contender is rhetorically cast an enemy. "Perceptions of the enemy,

distinctions between enemies, cannot therefore be treated as an epiphenom-
enon of the distinction between theological and political, but are instead constitutive of it" (Anidjar 2003, 32). In Colombian governmentality there has
been a repetition and reenactment, in different historical moments, of the
rhetorical and physical elimination of opposition. This is partially because the
idea of the political contender is based on a notion of alterity as inimical to the
state. Let us explore this relation between a politico-theological understanding of alterity and political opposition to government in relation to Caro, the
foundational ideologue of the nation.

Caro's argumentative style has been characterized as highly authoritarian
(Sierra Mejía 2002; León Gómez 2002). Adolfo León Gómez has analyzed
how such authoritarianism rested on a series of rhetorical moves, several of
which we can see in many of the passages quoted in this chapter: his prefer-
ence for a polemical style "that has a practical interest in debilitating the argument of the adversary, or even in destroying it" (León Gómez 2002, 155), the
ridiculing of his opponents, the denunciation of his opponents' claims as fal-
lacies while his own arguments are presented as indisputable truths, the use of
illustrations (examples) that are not presented as part of the argument but as
pragmatic support for his argument (2002). Deas ([1992] 2006) also mentions
the invocation of linguistic authority as a means Caro used to silence his opponents and his tendency to establish a truth for all times (past, present, and
future). In such a rhetorical style "the textual divisions that separate doctrine
from polemics and theological from political treatises, the historical division
that distinguishes between exegesis and practice . . . is rendered porous" (Anidjar 2003, 33). The long-lasting hegemony of conservatism and grammarians
at the beginning of the twentieth century made this discursive style into a
governmental practice, granting it enormous political power through its translation into the law and into the administration of the public sphere.

For Caro, as for many of his fellow conservative colleagues, Catholic religion guaranteed the social contract. The installation of religion as a political
theology of the state depended on the relation between the type of rhetoric he
deployed and the rhetoric of the law. As he argued in the statement he wrote
for the debate of the constitutional approval of Article 35 of the 1886 Constitution, Catholicism is Colombia's religion "because it brought civilization" to
the country. For him, "Catholic religion was our parents' religion, is ours, and
will be the only possible one for our children. Either that one or none" (Caro
[1886] 1980a, 1044). Tradition here emerges not to counteract the woes of
modernity (as in Isaacs's denunciation of the corrupting influence of modern trade on the purity of indigenous languages) but as the politico-religious

guarantee of a stable (and unchangeable) political future through solid intellectual argument. The notion of Colombia as a confessional state is based on the relation between temporality/culture (as tradition), religion, and language as a political guarantee of stability. This belief is expressed not only by Caro but is found in the writings of most of his conservative friends.[22] It is fundamental to understand Caro's appraisal of the relation between science and language.

Let us recall that he stated that the only reason for a scientifically practiced history is that of faith. Caro challenges the demonstrability of experimental scientific authority by stating that such a valorization of reason is based on the *belief* of reason as autonomous: "Science founds its assertion in demonstrations and its demonstrations in beliefs. . . . Rejecting the infallibility of principles and accepting that of demonstrations you want a science that does not exist and cannot exist. You cannot even know or envisage what true science is" (quoted by León Gómez 2004, 177–78). Ultimately, the scientific interpretation of reason as truth and the religious interpretation of science as based on faith are both rooted in belief. Caro cleverly unveils the problematic claims of scientific purification as autonomous reason. He does so in the name of replacing it with religious purification. In this way, he proclaims that both secular epistemic authority and religious authority are based on the power of truth that each grants to either reason or faith to enact the differentiation between society and nature. Here we see how the survival of religious ideals persists in science (Latour 1993; Stengers 1997). As stated by Whitehead, the "demand for an intellectual justification of brute experience has also been the motive power in the advance of European science. In this sense scientific interest is only a variant form of religious interest. Any survey of the scientific devotion to 'truth' as an ideal, will confirm this statement" ([1927/28] 1978, 15–16). Caro's astute move in the above passage is precisely that he recognizes the similarity of method (the search for unquestionable truth) even though he grants authority only to faith. Ultimately then, although the polemic between Caro and Isaacs positions them in very different camps regarding a politics of recognition of indigenous people, the similarity in the purificatory practices of science and religion reveal the adherence to a "devotion to truth" that links the scientific and the religious. But, "what is distinctive about secularism is that it presupposes new concepts of 'religion,' 'ethics,' and 'politics,' and new imperatives associated with them" (Asad 2003, 2). The consequence of this, as Foucault so clearly shows, is that the truth of science, the truth of religion and the truth of the state are differentially distributed in a Catholic state than in a secular one. Here it also implies the juridical deployment of the difference between myth (as characteristic of indigenous peoples) and religion (as characteristic of the state).

The "invention of world religions" was one of the key developments of the nineteenth century which resulted in the consequent division of rational peoples as worthy of religions from primitive peoples as practitioners of mythology (Masuzawa 2005). In Europe such a transformation was enabled by the perception by modern central-northern Europeans of the nineteenth century that religion was either disappearing or "becoming circumscribed in such a way that it was finally discernible as a distinct, and limited, phenomenon" (19). But what we have here is the contrary. There is an active deployment of religion in every dimension of the public sphere. In Caro, the interrelationship between grammar, faith, and science was meant to guarantee the temporal stability of the nation through their mutual purificatory practices. Thus the separation between indigenous peoples who professed "confused and fantastic remembrances of their origins" could not be done through the acknowledgment of a secular ascription of such religious practices to an "other" (as Isaacs tries to do in acknowledging their myths). In Caro, such people do not even have mythology, only "confused remembrances." For him, such an absence of intellectuality cum faith renders indigenous people as lacking one of the fundamental attributes of the juridical person. Such difference needs to be eliminated through the development of the proper intellectuality, which politically would guarantee bringing indigenous peoples into proper personhood, proper forms of devotion and into the rule of law. I believe the ideas Caro developed in his polemic with Isaacs were crucial in implementing the laws that determined the politics of indigeneity in the country. A few years after this polemic in 1892, Article 2 of Law 72 disposed that indigenous people were to come under the legal authority of missionaries:

> The government will regulate into law in accordance with ecclesiastical authority everything conducive to the good development of the Missions and will be able to delegate on the Missionaries extraordinary faculties in order to exercise civil, penal and judicial authority over its catechumens over which the action of national laws is suspended until the moment when, having left the savage state, they are in capacity of being governed by it. (In Sáchica 1991, 174)

The full import of a faith-based politics of listening to indigenous languages thus implies the reinscription of the politics of conversion as what guided the juridical sphere in indigenous affairs in Colombia, a politics that would prevail in good measure until the rise of indigenous social movements between the 1960s and the 1980s.[23]

The narrative of suspicion that Caro built around Isaacs rested on the "purification of listening" (Schmidt 2000) through its deauthorization as a scientific method. But its deauthorization did not relegate religious hearing, as in other historical moments or traditions, to the site of superstition and religious fantasy, but rather to that of secular poetry and sensibility. Throughout his text, Caro is willing to value Isaacs as a poet as when he values the poetic writing in his descriptions of indigenous myths. It is as if Caro wishes Isaacs's return to the author of *María,* the historical moment in which both were on the same political side. For Caro faith is part of the rule of law because it is inscribed in divine authority, not because it is revealed through, say, the place of the senses in enacting mystic communion. The ear of the confessional state, then, is prescriptive. One could say that the model of the confessional state was the Inquisition, in that it provided a template for the development of the relation between juridical development and the (painful) extraction of confession as that which established the authority between governmentality and the political subject (Asad 1993). As in the missionary endeavor, the objective was to extract from the subject the admission of an erroneous (sacred) practice. Indigenous persons' incommensurable form of life had to be brought into *policía cristiana* to eventually attain their "intellectualization," in order to redeem them from the state of savages and produce them as juridically commensurable to the state. The acoustics of the confessional state, then, are those of a prescriptive authority based on the admission of sin and the elimination of difference. An acoustic biopolitics is implemented through the full weight of the rule of law.

In the final decades of the nineteenth century several disciplinary developments, such as ethnography, psychoanalysis, medicine, communications, and acoustics, began to coalesce around the increased use of the ear for research (Sterne 2003). The questions raised by such acknowledgment of the ear's central role in knowledge constitution led to fundamental transformations of the assumptions about the relation between language and the ear in disciplines such as linguistics, medicine, and psychoanalysis. The need of the *letrados* to control the processes of language purification and their relation to power enacted a simultaneous purification of the practices of the ear. It is notable that Isaacs and Uricoechea, the two scholars, who were sensitive to a poetics and politics of hearing that potentially questioned a conservative interpretation of the lettered city's politico-theological relation to indigeneity, developed their theories under different forms of exile. As is evident in this chapter, and as is well known, the "American perspectives produced by some other best thinkers of the region are not repeated by its ruling classes, which constitutes one of the profound dramas of the continent" (Zalamea 2009, 13). Rather than marginal, the indigenous emerges here as centrally constitutive of the region's particular mode of perception and definition of itself, even if drastically excluded by the biopolitics of the nation-state. To be sure, the role of indigenous peoples is not to prescribe the limit of our own folk tradition of modernity (Viveiros de Castro 2010). But their recurring presence in a political space that repeatedly tries to do away with them redefines the way the very constitution of such a space needs to take stock of the indigenous repeatedly across history, in order to redefine itself intellectually and politically (Rivera Cusicanqui 2010). The dialectic between different regimes of the ear in the modes of relation between indigenous and nonindigenous persons has been crucial in such a history.

"We call *words of dubious orthography* those that can be wrongly written by a person who, in writing, follows no guide but the ear," wrote educator, philologist, and president of Colombia (1900–1904) José Manuel Marroquín (1827–1908) in his *Tratados de ortología y ortografía* (Treatises on orthology and orthography) (emphasis in the original) ([1869] 1874, 15). Toward the end of the nineteenth century throughout Latin America, knowledge acquired through the ear became increasingly suspect, giving rise to a grammaticalization of the voice with the institutionalized deployment of ever more formalized ideas about appropriate forms of vocality. The idea of "having a voice" as a metonym for a political subject and political participation developed alongside notions of sovereignty and liberty forged through the struggles of independence, the civil wars that followed in their immediate aftermath, and the rise of governmentality in the early postcolonial moment. Crucial to the formation of ideas of governmentality were the technologies of the written word: the increased importance of the circulation of ideas through print (Ramos [1989] 2003; Silva 2003), the transformation of jurisprudence from oral to written transactions of power (Gaitán Bohórquez 2012), and the consolidation of a new administrative and bureaucratic class of the independent nations with a need for a formal education to produce such a class—what Angel Rama called *la ciudad escrituraria* (the written city) (Rama [1984] 1996).

But the type of voice that metonymically embodied a political subject also had to be constituted through specific dispositives. By the end of the nineteenth century pedagogy became central to a project of optimizing voice as a disciplined knowledge. This would not only yield an appropriate idea of the person for the pastoral care required by governmentality but also, by the early twentieth century, give rise to an aesthesis of the vocal crucial to a style of folkloristics centered on an enlightened notion of the ear (Schmidt 2000) determined by the techniques of inscription of the lettered word as the proper model for the popular. Orthography and orthology (the discipline of correct pronunciation), etymology, and eloquence were technologies that shaped ideals about the popular in the voice. Through them, a series of "fantasies of magical omnipotence" (Connor 2006) were ascribed to the capacities of the voice to shape the political subject and the people. A virile vocality (Connor 2006) became a condition of the juridical-economic sphere and an idealized prescriptive vocal expressivity became associated to folkloristics. The relation between both shaped the political economy of the popular.

It has been argued that "orality" represents the disorder of the popular and is opposed to the lettered city. In this formulation, orality is one of those seemingly self-explanatory key terms that raises the issue of "why certain ideas can be both so vague and so well known, so easy to make concrete and so likely to engender haze" (Rancière [1998] 2011, 30). Despite repeated deconstructions the term *orality* maintains as self-evident, timeless notions of the person, of aesthetics, and of communication (Sterne 2011). Alternate terms such as "verbal art" (Bauman 1984), used in ethnopoetics and sociolinguistics, fields of study such as linguistic metapragmatics (Silverstein 1993), or different schools of "performance theory" have been largely developed as a broad response to problems underlying the idea of orality. But what is significant is that despite these critiques, which have been important for transforming the idea of vocalization in fields such as linguistics, anthropology, and studies in performing arts, the more widespread use of orality shows a political resilience that seems to resist such deconstructions. This is perhaps because of the central role it has played in the formation of modern political philosophy.

The relationship between spoken or sung language and tradition, that is, the sound of the voice and a certain conception of time, is central to the formation of the idea of modernity itself (Rama [1984] 2007; Ramos [1989] 2003; Bauman and Briggs 2003). As a central defining element of notions of community, folklore, and communication, orality partakes "of the metaphysical character of modern political philosophy revealed in its tendency to identify the sense of the big words of politics with their most immediately evident meaning" (Esposito [1998] 2009, 11). One of the underlying reasons for the endurance of the term despite repeated deconstructions is the way it is grounded in a political "theology of sound" (Sterne 2011) central to conceptions of the voice underpinning modern governmentality. In what follows I explore a particular history of orality seeking to understand how the voice came to be metaphysically imbued by certain notions central to modern political theology.

According to Andrés Bello, "the grammar of a language is the art of *speaking* it correctly, that is, according to good use, exemplified by educated people" (emphasis mine; Bello edited by Cuervo 1905, 1). Bello's definition is pledged to the performative distinctions of correct speech, and, as a "pedagogical dispositive" it comes to occupy "an intermediate space between (irreflexive) speech and the rationality of writing" (Ramos [1989] 2003, 68). Late nineteenth-century philologists in Colombia, particularly Miguel Antonio Caro, Rufino José Cuervo, and José Manuel Marroquín, intervened in this intermediary space in order to manage not only the relation between speech and the written word, but also the relation between articulation and timbre as the key constitutive elements of a proper voice. The idea of orality that was the basis for early twentieth-century politics and folkloristics emerged from such a vocal notion of grammar and functions simultaneously as a term of political philosophy and of an aesthesis of the popular. By orality, then, instead of an opposition to the lettered world, I designate a historical mode of *audibility* of the voice linked to the rise of its grammaticalization (as it was then understood) and to the concomitant silencing of untamed vocalities that refused to submit to such grammatical acoustics.

Orality/untamed vocality form a mutually constitutive pair, but while one side of the term is assigned a supposed conceptual transparency (orality deals with the spoken word), the other (untamed vocality) is actually subsumed under the concept of orality. Underlying both the concept of an untamed vocality and of grammar as a training in speech is the acknowledgment that voice is a "powerful emanation from the body" (Rosolato 1974, 76). Orality is that discipline which while recognizing the powerful relationship between the ear and the voice as a central element of animated, living beings, rendered it into a particular politics of differentiation between the human and nonhuman. Orality was not what, in the late nineteenth century, named the multiplicity and singularity of different vocalities but rather what disciplined the production and perception of the human voice. This notion was developed in relation to ideas about appropriate forms of vocality and saying (*el saber decir*; Ramos [1989] 2003) crucial to late nineteenth-century politics, forms of valuing the vocal arts, and ideas about music derived from the voice. As such it is a historical mode of audibility of the voice produced, on the one hand, by the grammaticalization of the relation between the spoken, sung, and written word, and on the other, by the systematization of popular expressivity (see also chapter 2). In what follows, I intend to explore the historical process that led to the emergence of orality as a concept in late nineteenth-century Colombia through ideas and pedagogical practices regarding eloquence, etymology and orthography.

The processes of political independence from the colony generated tremendous anxiety among Creole elites about the possibility of fragmentation of the Spanish language (Ramos [1989] 2003). No longer did a single sovereign power reign any longer over all the Americas to guarantee its cohesion. This transitional moment raised the specter of diversification of pronunciation across the vast, unconnected territories of the Americas and the potential development of different, unintelligible new languages. This gave rise to an "alarmist tradition" in Hispanic linguistics, shaped by perceived threats to Spanish in the post-independence period (Del Valle 2002a). The work of Venezuelan grammarian Andrés Bello was highly influential in shaping this project (Ramos [1989] 2003). Colombian philologists Rufino José Cuervo and Miguel Antonio Caro studied Bello's *Gramática de la Lengua Castellana* while in school, cited his works frequently, and dedicated repeated and extensive books to the revision of Bello's work. Cuervo revised and indexed Bello's grammar on multiple occasions in his *Notas a la gramática de la lengua castellana de D. Andrés Bello e índice alfabético de la misma obra* (Notes to the grammar of the Castilian tongue by D. Andrés Bello and alphabetic index of the same work). Caro did the same with Bello's *Ortología y métrica de la lengua castellana* (Orthology and metrics of the Castilian tongue) (printed in Bogotá in 1862 and 1872) in his *Notas a la ortología y métrica de Don Andrés Bello* (Notes to the orthology and metrics of Don Andrés Bello) (1882). Several passages in their work are taken nearly verbatim from Bello, a testament to their reverence for him rather than of plagiarism.

In the hands of philologists with presidential power, such as Caro and Marroquín, Caro's and Cuervo's obsessive re-citation and "correction" of their master's work turned his pedagogical–political proposals into linguistic dogma, into pedagogical method, into the argumentative artistry of jurisprudence, and into the aesthesis of the folkloric. Bello, Caro, and Marroquín were all central figures in the development of universities, the legal sphere, and, as has been mentioned in previous chapters, even presidents of their nations (in the case of Caro and Marroquín). A central aspect of the emergence of the audibility of orality was determined by this capacity for its institutional deployment central to the formation of a sense of Latin Americanism (De la Campa 1999) shaped by the exchange of publications and by a large epistolary that evidence the formation of a shared field of knowledge and cultural policy (Miller and Yudice 2002).

Acknowledging the relation between the expressive aesthesis of the voice in the popular and the juridical omnipotence of a virile voice for the sovereign subject became a central aspect of this project. This implied taking into account the political question posed by fact that different dimensions of the voice were perceived as shared by humans and nonhumans. As stated by Miguel Antonio Caro:

> Not only man, but also animals have the gift of the voice, and use a certain sonorous language, but it is *inarticulate*. Man uses a language similar [to that of an animal] when he cries or screams of happiness or terror, when he whines or complains, ultimately, when he emits voices without speaking (*cuando vocea sin hablar*). This is the inarticulate language common to man and animal that expresses faculties that are also common to one and to the other: the sensitive faculty that consists in expressing pleasure and pain and the estimative (*estimativa*) faculty through which the animal appreciates (and man too) instinctively, without the use of reason, what is convenient or repugnant to his physical nature. Thus if a hurt or wounded animal flees howling, it expresses an act of its sensitive faculty—a pain; if it gives voice to ask for food (*si da voces para pedir alimento*), to announce danger or something similar, it expresses an act of its estimative faculty. . . . *Articulate*, human language expresses acts of a superior faculty to those two previously mentioned and peculiar to rational beings—the *intellectual* faculty (*la facultad intelectiva*). Words, elements of language, represent ideas, forms of thought. If a man in pain whimpers, showing his pain, he has made use of inarticulate animal language. But if he wants to express that same pain through words . . . in that case, he expresses *directly* what he *thinks* about what he feels and *indirectly* the *sensation* he is experiencing. (Emphasis in the original) (Caro [1881] 1980, 448–49)

For Caro, the voice, then, manifested the animal dimension of the human in terms of expression of sensations and of instinctive reactions to physical survival. Here what is emphasized is the voice as "the body's greatest power of emanation" (Rosolato 1974, 74), as "the flesh of the soul" that expresses its sensations (Dolar 2006, 71), as that "which ties the signifier to the body" (59). In this conceptualization voice "is what holds bodies and languages together" (60) and voice is seen as emanating from a vocalic entity to an outside hearer. In this latter sense it is always "in transit" *between* beings (Connor 2000, 23) and thus central to the constitution of ideas of relationality. Conceived as simultaneously holding together bodily and acoustic expressivity and as migrating between a speaking entity and another hearing one, voice has the potential of appearing to our ears as "the undeniable evidence of will or intention" (Connor 2000, 23). Such evidence of will or intention was key to an enlightened, rational, articulate notion of the voice that contrasts with the metaphysical intentions of oracular voices (Schmidt 2000), to ones attuned to the subtleties of interspecies vocal exchange as when songs are taught to humans by animals (Seeger 1987) or to different intensities of the human as when envoicing is done through musical instruments (Hill and Chaumeil 2011) or as when a different being manifests vocally in the body of a person that is "possessed" by another entity. But, following a long Aristotelian metaphysical tradition, what made the human voice different from that of animals for Caro was its articulate, rational character expressed in the relation between words, and ideas as crucial to the constitution of personhood.[1] In this prevailing lineage of Western metaphysics "man presupposes a *zoè* that also transforms his voice into an articulate language" (Ludueña 2010, 33). Cultivating the voice was part of a project of training the dubious knowledge of the ear, "improving the senses and building trust in their discriminatory power" (Schmidt 2000, 137) in order to produce the supplementation necessary for the fantasy of life associated to this desired *zoè*:

> The voice is so saturated by the anxious dream of our "life," because it is itself one of the most important components of that will-to-life. The phantasm of the "living voice" is the principal carrier of our hallucination of life. It is subject to a paradoxical vital economy. While drawing on the body for its force, and therefore subject to the vicissitudes of the body, it is nevertheless imagined to have the power to radiate new life back to the body from which it emanates. But, as a surrogate or supplement, the voice is also itself in need of supplementation—hence the anxious regimes of voice cultivation, nurture, hygiene and healing which have multiplied since the end of the eighteenth century. (Connor 2006, online document)

The voice's vital economy had to be supplemented so its enormous force could become articulate. The question emerges then, what is an articulate vocality? In Caro, articulate vocality is mediated by the feedback between two notions of linguistic transparency: a phoné that can be clearly heard and understood due to clear pronunciation and resonant timbre, and a clear enunciation of concepts through the cultivation of the relation between rhetoric and intellect. The philologists had to take charge of the zoopolitical voice of the population by intervening in the relation between the signifier and the signified. And they did so, not by excluding voice from written language thereby making it into its opposite (as is supposed in ideas of orality) but rather by immunizing it (Esposito [1998] 2009), that is, acknowledging its powerful expressivity yet developing the means to protect the self from its unintended outcomes. Vocal immunity uses the fear of voice's intrinsic potential for manifesting an incoherent or otherwise undesirable form of the self to produce a vocally articulate one, grammaticalizes the voice through the rules of writing while purporting to speak in the name of "people's" audible vocality, and curtails the dubious ear's reception of the voice by training it to distinguish and parcel out the uses and functions of proper and improper voices amongst different peoples.

Immunity is created in response to (or against) the very materials of which a virus is made to prevent its full-blown presence and effects. It is not exclusion but rather a particular form of "conjunction" (Ludueña 2010) and incorporation that determines how what is named in a term is prevented from growing into full expression. To immunize the voice is to use the power of the voice to obfuscate its modes of presence in order to prevent uses that are understood as undesirable. Inoculation processes were means by which "new technologies of population control, record-keeping, and classification . . . that projected . . . medicine's agency in the ordering of a rationalized social bond" were implemented in the nineteenth century in Latin America and the Caribbean (Ramos 1994, 187). This reference to medicine acknowledges not only the actual rise to power of the profession but of its procedures as transversal techniques that served as conceptual models for other fields (187). The projects for immunizing the voice took place through specific "anthropotechnologies," that sought "to fabricate the human as *ex-tasis* of the animal condition" (2010, 11).

The immunization of the voice through specific anthropotechnologies gave rise to two intertwined notions of "the voice of the people." As stated above, metonymically, the notion of "voice" has been used in the West to represent the capacity of a particular political subject's participation in the public sphere. Such a notion of voice implies the constitution of a political subject as one who "has" an identity that through communication acquires the possibility of transparent political participation through representation. The notion of the person that underlies it implies a move away from "surfaces" of acoustic/sensorial vocal production and perception to generalizable grammatical "models" (Deleuze 1969) about the voice in order to produce a clear identity. Here we have the desired communicative transparency of the autonomous individual, who at the same time has ethical "depth" and is expressing an interior will central to the formation of "the people" as a political subject. Such a political subject, in turn, exists by virtue of ceding his (and it was his at this time) possibility of representation to the sovereign, linking political subjectivity to submission (Esposito [1998] 2009).

Yet theories of vocality emerged not only to highlight the individual's political participation but as an aesthesis of vocal communion in folklore that represented ideals of unity, spontaneity, adherence to the past, heightened sensorial perception and emotional expressivity, and anonymity. This was a voice characterized by lack of authority (without an author) and by a lack of creativity (a voice that reproduced the past in the present). At the same time, this voice was meant to carry, in the aesthesis of popular expression, what had been stripped away from the politically representative voice. Here, vocal immunization takes a different form: in the name of an aesthesis of expressivity and affect, such expressivity is impersonalized into the spontaneity said to be typical of the gift-giving community. It thus becomes a "third person" politics (Esposito 2009a) grafted onto a theory of affect, spontaneity and idealized benevolence that characterized a whole people (the folk) and their expressions rather than a person.

Such a double model of immunity created a spectral figuration of the popular in the political. A contrast between the popular as politically representative yet aesthetically deplored and the popular as aesthetically significant yet impersonal was crucial to the formation of the political philosophy that came to constitute the idea of "orality" (Martín-Barbero [1987] 2001). The notion of orality emerged in the spectral relation *between* the two immunitary paradigms of vocality: one that seeks to present the person as political subject while masking the voice in the name of political participation, and one that seeks to present the expressive subject while masking rationality and individual creativity in the name of communal expressivity and sensorial intensity. In this project, the untamed voice, understood as "an index of bodily exaltation" (Rosolato 1974, 76), remained as an infrasound of what could potentially happen to the human animal if it were unleashed. The very pedagogy to tame the voice was based on creating vocalic routines that masked the "animal becoming" (Deleuze and Guattari [1987] 2011) of vocality's power present in both vocal paradigms of the popular. In what follows I want to explore how the anthropotechnologies of eloquence, etymology, and orthography, as historically specific projects of acoustic inscription, tried to immunize the voice by seeking to contain its acknowledged powers.

Throughout the nineteenth century, the dialectic between an appropriate elocution and Latin American (mis)pronunciation of Spanish was a subject of repeated theorization by lettered men who were actively involved with the arts of government in countries where the boundaries between barbarism and civilization were perceived as being way too flimsy (Ramos [1989] 2003). But as expressed in the previous chapter, it is not until the late nineteenth century with the Regeneration and the rule of the conservative Catholic government that such a question was actively deployed through a pedagogical and juridical dispositive in Colombia. In this process, apparently contradictory ideals of vocality had to be brought together. The problem was not only that of distributing the valorization of the aesthetic aspect of the people in an "orality" subservient to the mispronunciation of "popular poetry" while simultaneously upholding and grafting a metonymic idealization of the voice of the people as a sociopolitical strategy (Martín-Barbero [1987] 2001). What was at stake was a zoopolitics of the person that is not simply resolved in the distribution of the valorization of orality between aesthetics and politics because both those who have and who do not have elocution, understood as a proper mode of speech devoid of mispronunciation, are also supposed to "have" the voice of the subject to the sovereign. Thus, the fundamental zoopolitical problem lies not in spoken language per se, but in the relation between different conceptions of the voice in the constitution of the relationship between the individual and the collective for the notions of "the people" central to ideals of community and of sovereignty. In the new, republican governmental situation, the problems of the relationship between "the people" and (mis)pronunciation were clear to Miguel Antonio Caro:

> As with the notion of "the public" (*el público*), one can ask of her [pronunciation], where does one see it? Where does it live? And it is no less mobile and multiple than the public. The rich man does not pronounce like the poor man, the person from city A like that from town B, the person who was born in one's same year like he who delayed his arrival into the world for ten, twenty years. To give the scepter of Orthography to Pronunciation, is like giving it to the one-hundred-headed monster. . . and just as once the sovereignty of the people has been proclaimed, the most insignificant local person comes to be called a people, once the sovereignty of pronunciation is proclaimed, the most obscure bad habit in speech becomes pronunciation. What a tower of Babel! (Caro [1867] 1980, 356–57)

One of the problems underlying the unintelligibility produced by the multiplicity and mobility of pronunciation is that of the relationship between reality and representation: "is it even possible that orthography is a system that represents pronunciation with fidelity and precision?" asks Caro in the sentence prior to the passage quoted above (356). Since the answer is obviously negative, the relationship between acoustic reality as it manifests in the voice of a person and acoustic representation as it manifests in orthography had to be brought into a positive relation through the control of multiplicity and mobility. But such control is not done through the production of a uniform, written grammar. For Caro, written texts do not make a language uniform, *"only oral tradition does"* (my emphasis) (Caro [1884] 1980, 354). However, "none of the grammar texts that children have teach them the pronunciation of letters" since they only teach them syntax (354). Thus, the problem of uniformity of language and uniformity of a people emerges as one of a pedagogical relation between pronunciation and orthography where the intervention has to happen first at the level of vocalization.

The question that Caro names by invoking Babel and the lack of intelligibility is one of the relationship between translation and survival (Derrida 2002). However, in the case of Caro the danger that looms in the background as a spectral future is not difference between tongues but the difference in pronunciation. The problem is to determine a process to decide which of all the possibilities of pronunciation is to survive in the relationship between orthography and orthology in such a situation (Ramos [1989] 2003). Since there is no "original proper pronunciation," a competent authority must determine which one is the most appropriate to serve as a model. The distinctions between phonetics, as the linguistic-physical discipline of vocal articulation, physiology as the science that aids the description of the physical production of tones, and philology, as a discipline concerned with heritage and kinship, begins to take shape in Latin America through a zoopolitics of the voice concerned with its eugenesis. Central to this scientific differentiation and its potential use in a eugenesic project of the voice is the distinction between voice, language, and eloquence produced by the art of elocution.

"The human voice is a current of air that becomes sonorous through the vibration of two 'vocal chords.' Voice, properly speaking, is the sonorous voice; when air is emitted without that sonority, it is only breath, murmur, and puff (*aliento, susurro y soplo*)" (Caro [1884] 1980, 385). Language, on the other hand, "is essentially a system of sounds" (383–84) but it can also be considered "as sound and sign" (Caro [1881] 1980, 446). Such a system consists of consonants and vowels which are "the physiological division of letters" (453). Vowels "are the sound of the voice" and consonants are "noise" (386). Such a "noise" is produced "by the modifications impressed on the current of air by diverse postures of the mouth, that is, by the diverse forms of contact of the organs—palate, tongue, teeth" (Caro [1884] 1980, 385). Vowels, on the other hand, "are essentially sounds that can be intoned or sung; musical letters. Consonants are not necessary for the person that sings, but nobody can sing without simultaneously producing a vowel" (Caro [1881] 1980, 454). Vowels are also "differentiated one from the other, as are the timbres of different musical instruments" (454).

Caro develops in these statements a philosophy of the voice in which vocal physiology, articulation, and intonation emerge as interrelated acoustic aspects that need to be trained. If language is what distinguishes humans from animals because it is "articulate" and expresses the "intellectual faculty" particular only to human beings, such intellectual faculty is manifested not only in syntax and semantics but explicitly in the acoustic. Physiology provides the elements for a refunctionalization of the vocal apparatus as something that can be scientifically understood and phonetics as the means of pedagogically implementing such understanding. This scientific relation is particularly important because the system of sounds that is crucial to language is arbitrary for several reasons: first, "anfibology," the lack of clarity in certain terms due to polisemy, is a trait of every language "due to the variety of significations that each word suffers" (Cuervo 1905, vii). Second, the figuration of language in written signs is also problematic because one single written sign corresponds to more than one sound. So, says Caro "figuration is sterile" (Caro [1884] 1980, 382–83). Third, the vocal organs "experience notable alterations by reason of races, climates, and even hereditary aspects. Neither the disposition of vocal organs is identical nor the special configuration that results from custom" (383). Thus nineteenth-century Americanist theories of the relation between humans and climate also determine the voice (see chapter 2 for a further elaboration of this theory). Such a tendency to disjuncture between sound and sign leave the vocal production of language dangerously close to an "inarticulate" animal voice in need of containment by elocution.

"Elocution," says Miguel Antonio Caro, "is the art of producing sounds, words and clauses with precision and propriety, with the adequate modulation and expression, when we speak or read. Voice is the instrument of elocution and language the essential form in which it is exercised" ([1881] 1980, 446). It has two elements: "grammatical phonetics" and "musical phonetics." The first deals with "the correct pronunciation of words . . . subdivided into vocalization and articulation" as well as the proper use of "punctuation and pauses." The musical aspect of elocution deals with "prosodic accent" and the "modulation of general phrases and periods, especially that of verses, with convenient tones and rhythms" (451). The analogy between music and grammar is possible because both are understood as systems based on the organization of sounds, a complementary poetics that expresses an intellectual aspect and the adequate cultivation of a sentiment. So the problem of pronunciation is not only pronunciation but also rhythm and tone. Elocution is a means to make the dubious ear, as manifested in the arbitrariness of the voice, reliable through the control of its prosodic and musical dimensions. The genteel refinement provoked by elocution created, in turn, an eloquent authority shared by lettered men who adequately cultivated their speech in the republic of letters (Ramos [1989] 2003, 62).

Eloquence had two objectives: "*to persuade* and *move*, to speak to understanding (*hablar al entendimiento*) and to touch the heart" (Caro [1881] 1980, 450). Says Caro, "Elocution has a grammatical and logical aspect that refers to ideas, and another expressive and musical one that refers to sentiments (*sentimientos*). *Sentiments*, we have said, and not *sensations*, because sensations is the name ordinarily and privately given to that genre of sensibility that is common to humans and animals. Sentiments are acts of an order of sensibility more exquisite and noble, characteristic of rational beings" (450). While each species has a characteristic inarticulate voice through which it expresses sensations (frogs croak, cows moo, and so on), man sings. "Articulate speech is not song, but it is broken and modulated with accents, tones and rhythms that imitate song in order to express the sentiments that animate us" (450). Such musicality is what guaranteed transcendence by differentiating between sentiment and sensation associating "idea and sentiment, idea and expression." Thus "words emerge from the human lip as from a musical instrument that exceeds all of us, 'with that ineffable vibration, that divine ardor that is born directly from the soul and that penetrates it so profoundly'" (Caro citing Coll y Vehí [1881] 1980, 451). While the power of the voice is immanent to every being and every being expresses sentience, the power of sentiment is only available to humans through the musicality of language. Elocution was the anthropotechnology that needed to be implemented in order to separate animal sensation from human sensibility acting upon the human biological substrate in order to produce an adequately eloquent person as one who transcends sensation through the cultivation of a musicality of the voice that renders sentiment "divine." The musicality of the voice gave a sonorous body, present only in eloquence, to the political theology of human transcendence. Voice had to be hominized through acoustic techniques that cultivated the relation between musical sensibility and grammatical rationality, both of them understood as dimensions of the sonorous aspects of vocality in language.

In this modern zoopolitics sentience "does not have a separate or separable existence from the subject who through it learns to know reality. Sentient life is rigorously limited and reduced to the simple psychic and sensorial knowledge *internal to a subject*" (Coccia 2011, 15). Thus voice is immunized by grafting it onto music and grammar, protecting it from its flimsy boundary with sentience, its centrality in the history of the sensuous (Dolar 2006) and its callings to humans' animal becomings. Conversely, it divests the manifestation of intentional sentience away from other entities in the world: "only the prohibition of intentional species allowed for the subject to coincide with thinking (and with thought) in all its forms" (Coccia 2011, 16). Thus, we get a folk cosmology in which only humans have a voice with intentional sentience as a condition to transcendence. This was a crucial move to create "paradigms of culture derived from notions of belief and conversion" in order to "conceive of culture in a theological mode, as a system of beliefs to which individuals adhere, so to speak, religiously" (Viveiros de Castro 2011, 12). Music is here believed to be the art form that gives speech its sound theological transcendence. So, the lettered city is less a devocalization of grammar because of the rise of the written city (*la ciudad escrituraria*) than an immunization of its vocality by bringing it into acoustic transcendence and into the secular realm of culture. The age of print is also the age of technologies that simultaneously rendered sonorous certain aspects of the voice through their careful cultivation while silencing others.

In bringing acoustic sentience into human intentionality, the grammarians reinscribed the conservative, classical regime of representation of the arts as a modern one. Rancière says that the system of representation of the arts that characterized classical theater "consisted less in formal rules than in their spirit, that is, in a particular idea of the relations between speech and action" ([1998] 2011, 44). The relation between speech and action in this regime was based on a certain deployment of verisimilitude in fiction. A type of person (a noble person, for example) was represented on stage through a type of speech corresponding to its character. Even though it was understood that in real life such correspondence was not necessarily precise or coherent, it was expected to be so on the stage: "The system of representation depends upon the equivalence between the act of representation and the affirmation of speech as action" and is guided by the "primacy of the speech-act" as the norm of the "edifice of representation" (48). In such a system "this ideal of efficacious speech in turn refers back to an art that is more than an art, that is, a manner of living, of dealing with human and divine affairs: rhetoric" (48). What Rancière calls rhetoric is what Caro called oratory, understood as the most sublime of the practices of eloquence. Accordingly, for Caro, such a regime in which the drama of the speech–act is the drama of the person, the utmost rendition of proper eloquence is that done by actors at the theater who express at the same time that they represent, establishing a proper equilibrium between sentiment and intellect (Caro [1881] 1980, 440–44).

Rhetoric, according to Rancière, is appropriate for "the age of revolutionary assemblies" ([1998] 2011, 49). But Latin America at the end of the nineteenth century is not the age of revolutionary assemblies. This is the moment after the wars of independence and the nineteenth-century civil wars, when revolution begins to give way to a rational constitution of governmentality guided by science and progress. This was accompanied by the transformation of natural history into science (Nieto Olarte 2007), of exploration into geography and engineering (Sánchez 1999), of the chronicler into the professional author (Ramos [1989] 2003) and of the dispersed collection of subaltern expressive practices into the canonic regimentation of the popular. This is a historical moment that *simultaneously* upholds conservative ideals of representation in spoken language and rhetoric while engaging in processes of modernization. Here eloquent speech is simultaneously the language of oratory and persuasion and "the silent speech of what does not speak in the language of words, of what makes words speak otherwise than as instruments of a discourse of persuasion or seduction: as symbols of the power of the Word, the power by which the Word becomes flesh" (Rancière [1998] 2011, 55). In Latin America and the Caribbean, "eloquence" names "an unequal development, in which a form of traditional authority (eloquence) is refunctionalized, even operating as an agent of the rationalization that would ultimately displace it" (Ramos [1989] 2003, 67). That is why "the concept of the modern episteme as fragmentation of general knowledge cannot be applied to the Latin American nineteenth century . . . we have to speak of an unequal modernization that surpasses (*desborda*) the categories of European historiography" (67). Eloquence understood as part of the representative regime, optimized in a politics of vocal pedagogy was to play a crucial role in the arts of good government, in education, and in jurisprudence as the performative sites through which such speech-acts were tied to ideals of progress through the constitution of a person appropriate to the scientific rational and economic labors it required (Ramos [1989] 2003). As such, reading out loud and reciting well, aspects of elocution, are "arts of imitation" that had to be taught in schools "in case that governments manage to entrust [the arts of good pronunciation] to experimented and well selected teachers" (Caro [1881] 1980, 443) but also brought into the sphere of science as that which deployed the invisible truth of the vocal, that truth available only to the expertise of the scientist, in the understanding of the relation between physiology and phonetics.

But it was not Caro who designed this pedagogical program. It was José Manuel Marroquín (1827–1908), president of Colombia from 1900 to 1904.

The most important pedagogical work on correct orthography in Colombia was probably José Manuel Marroquín's *Tratado Completo de Ortografía Castellana* (Complete Treatise on Castilian Orthography), which was originally written in 1858, had undergone four editions by 1866, and had subsequent editions after that date. This treatise was accompanied by another work, the *Tratado Completo de Ortología Castellana* (Complete treatise on Castilian orthology), which appeared for the first time in the fourth edition of the Treatise on Orthography (1874, iv). In the prologue to the fifth edition of the orthography treatise (or the second edition of the orthology manual, which was first published in 1869), the book is presented as distinct from others on the subject, for example, those of Andrés Bello. Unlike scientific studies of the topic meant for learned scholars, Marroquín's book was conceived as a practical pedagogical work to be used in schools, a manual for beginners to learn to write, and for teachers to teach orthography ([1869] 1874, v). It was meant to produce a standardization of the Spanish language following the rules of usage of orthography proposed by the Spanish Academy of Letters. This was a crucial work in the reeducation of perception, in procedures of how to refine the dubious ear to produce reasonable ways of hearing and pronunciation, thus producing an appropriately representative acoustics of speech. And the medium with which to do it was not just the written word but the models of vernacular oral poetry used as technologies of enlightened eloquence.

In this manual, orthology is defined as "that aspect of grammar that deals with the pronunciation of words," which consists of three aspects: "teachings regarding elemental sounds, regarding accents, and the rules for distinguishing the combination of vowels that form diphthongs and triphthongs from other combinations" ([1869] 1874, 9).[2] The main objective of this sonic guide is to provide the guidelines to create standards of pronunciation, and, on the way, standards of hearing—an enlightened technology of disenchantment and "realized emptiness" meant to empty the ear of its dubious hearing and produce a reasonable listener with powers of selective discernment, a sensory education meant to establish "the right habits of mind" (Schmidt 2000, 3). This pedagogical intention is also found in the treatise on orthography that follows the treatise on orthology. Marroquín defines the orthography of Spanish as "the art of representing according to written use, the sounds of which the words of the same tongue are composed" ([1869] 1874, 27). The book consists of an enumeration of the laws of orthography followed by a list of words of common usage that have "dubious orthography," arranged both in alphabetical order and in verse form, following the structure of the popular four verse coplas. "The teachings on orthography would be very incomplete if we had not filled with long catalogues of words [*voces,* literally voices] the void left by rules. We have taken into account the difficulty that beginners would have in memorizing them, and in order to avoid it, we have rendered all of these catalogues in verse. As we have been taught by experience, this makes them easy to memorize. It is enough for children to listen to these catalogues in verse being read to them every day for six months or one year, in order to have them indelibly sketched on their memory" ([1869] 1874, 23). Schoolchildren were taught to memorize the spelling catalogue in verse. My mother, for instance, still uses the 1952 edition of her catalogue of words for spelling, and I remember hearing my grandmother teaching me how to spell by playfully reciting the verses of those words of dubious orthography which use the z, as one who recites a piece of oral folklore:

> Con zeta se escriben almizcle, vergüenza,
> Hozar, despanzurra, bizcocho, azafrán,
> Azufre, bizarro, calzones i trenza,
> Coraza, lechuza, durazno, alazán.

> With z we write musk, embarrassment,
> Rooting, breakage, cake, saffron,
> Sulphur, bizarre, pants and braid,
> Armor, owl, peach, sorrel.

Rendered as popular poetry, rhythm and intonation were also a crucial aspect of this verbal art of orthographic pronunciation. Colombian philologists spent an enormous amount of energy composing works that would create a proper relation between orality and writing not only training the mouth to pronounce properly and intone with perfect rhythm, and the hand to write with good orthography, but also to produce "the trained ear with its carefully acquired perceptions" and a "listener with well cultivated powers of selective attention" (Schmidt 2000, 3). If the ear produced dubious knowledge, all one had to do was train it into acoustic reason. But the problem remains: in the many variants of pronunciation of a single word, how does one select the appropriate one?

ETYMOLOGY

Due to its tendency to change across generations and territories, language was considered "a living body" (*un cuerpo viviente*) (Bello 1905, viii) and, like a living body, it was characterized by different "life epochs" (Cuervo [1914] 1987,

23). If one of the problems of proper language pronunciation and orthography was multiplicity, the other was change over time. Etymological techniques emerged as the means to control language's tendency, as a "living body," toward diversification over time, by selectively determining the correct origin of a word in order to authorize its proper use in the present. In the hands of late nineteenth-century Colombian grammarians, etymology was used as a "strategic maneuver of territorialization and temporalization" (Povinelli 2011, 16) for determining proper heritage through lexical comparison in order to weed out the vices of ill-advised linguistic transformations. Etymology thus became a technique for a eugenesis of the tongue tied to a project of national sovereignty. If eloquence turned the multiple into one form of speech (and one people), etymology provided the means to arrive at the definition of what or who that one should be. No other philologist in Latin America dedicated himself with such obsessive detail and fervor to the labor of etymological selection than Rufino José Cuervo (Bogotá 1844–Paris 1911).

Cuervo grew up in Bogotá and attended the same school as his future colleague and close friend Miguel Antonio Caro. His father died while he was still young and his brother Angel took care of the family, which he sustained economically through the proceeds of a beer brewing company he had founded. In his youth, Cuervo worked in this company collecting receipts from taverns and bars, and it is thought that these are probably the places where he heard much of the local language he annotated (Rojas 2004; Vallejo 2007). He and his brother first traveled to Europe in 1878, where they spent a year visiting European philologists and acquiring the most recent books on the subject (Rojas 2004). They returned to Bogotá in 1879, but only briefly. In 1882 they sold the brewery and moved to Paris where Cuervo worked on his encyclopedic dictionaries for the rest of his life.

Like much philological work of the period, Cuervo's oeuvre is monumental in purpose and encyclopedic in scope. He dedicated his lifework to the writing of annotated dictionaries of the Spanish language. His *Apuntaciones críticas sobre el lenguaje bogotano* (Critical Notes on the Language of Bogotá) appeared for the first time in 1867 and was subsequently published and revised six more times during his lifetime.[3] His other work is his *Diccionario de construcción y régimen de la lengua castellana* (Dictionary of Construction and Regimentation of the Castilian Tongue), of which he only managed to publish two volumes, in 1886 and 1893 respectively, covering the letters A through D. He also edited, annotated, and "corrected" Andrés Bello's reeditions of *Gramática de la lengua castellana*. Cuervo was obsessively concerned with the search for the perfect form, with the continual weeding out of mistaken citations, wrong words, and ill-advised linguistic turns, carefully honing the philological authority to do so through detailed, repetitive attention to every single dimension of lexical comparison. He treated authoritative texts (such as Andrés Bello's *Gramática* or his own work) with the same scrupulous demand for the appropriate form as he treated lexical history. He spent his life continually correcting, revising and reediting his own work and that of his admired model figure, Andrés Bello. Throughout his life, he applied to his own work and that of Bello the revisionist attention that he learned from the comparative lexicographical method that was the basis for his etymological thinking. Such obsessive revisionism and attention to correctness made him a key figure of the "alarmist tradition" of Hispanic linguistics (Del Valle 2002).[4]

For Cuervo the introduction of changes in pronunciation, sounds, grammatical structures and word usage happened through the presence of provincialisms, neologisms, and archaisms. These are respectively defined as new words that took form in the localization of language, newly coined ones, and those that persisted in the Americas even though they might have disappeared in Spain. Especially troublesome, according to Bello, was the incorporation of "neologisms" that "alter the structure of language and tend to convert it into a multitude of irregular, licentious, and barbarous languages, embryos of future languages that through a long elaboration would produce in America what happened in Europe during the dark period of corruption of Latin" (Bello 1905, vii–viii). Cuervo also warned against that "promiscuous crowd of voices and constructions of which the speech of a people is composed" (*la promiscua muchedumbre de voces y construcciones de que se compone el habla de un pueblo*) (1886, xxxi). Fear and alarm of uncontrolled differentiation across time expressed through metaphors of sexual contact and reproduction permeated these philologists' discourse on language change, seen in the idea of linguistic gestation, uncharted mixed descent, licentiousness, and promiscuity found in the above quotes. Such "promiscuity" needed to be transformed into "a hierarchy of social power tied to the prevalence of genealogical familial authority" (Irigaray 1996, 13–14), especially due to the key role of language in mediating the relationship between the family and the nation: "Nothing, according to us, symbolizes the Fatherland as clearly as language (*la lengua*): in it is enfleshed (*se encarna*) that which is sweetest and dearest to the individual and the family, from the sentence learned from the maternal lip and the stories told in the love of the hearth to the desolation brought by the death of parents and the dimming of the home" (Cuervo [1914] 1987, 6).

Bringing linguistic promiscuity into appropriate genealogical descent from the "maternal lip" that exists to the "love of the hearth" is to move language from an unpoliced reproduction to one of maternal love intended for the existence of the fatherland. Since language change was inevitable, the etymologist was needed in order to provide "the authority that decides when to cede to the invasion of a novelty" (Caro [1867] 1980, 358). The Colombian etymologists were less invested in a discourse of disappearance of subaltern peoples and the consequent need to preserve their expressive culture that characterized the emergence of transcription (see chapter 1) and mechanical recording techniques (Sterne 2003) and more invested in controlling the process of transformation of expressive culture according to their own principles of heritage. This happened through the negation of transformation as a source of potential multiplicity and the creation of a regime of general equivalence for subaltern expressive culture. Etymology's methods and assumptions provided a means of appropriate management of heritage crucial to the consolidation of patrimony as a national paterfamilias. The etymologists developed patrimonial eugenesic techniques to protect a selective repertoire of appropriate expressive traits to guarantee the healthy survival of the fittest expressive forms by pretending to control the means of descent. They pretended to alter the mode of transference from one generation to another through their intervention in the history of words. Through this, they produced a zoopolitics of the voice that transformed the idea of the person from one that was determined by the relation between race and sexuality in the politics of blood purity during the colonial era to one determined by the relationship between the family and the nation during republican times. This was also a move from the slave as the economic figure of the person for colonial times to the autonomous individual (or "the people") as a productive and articulate labor force for modern times (Ramos 1994). This masculinization of descent as a temporalizing anthropotechnology of the state was a central aspect of the Latin American idea of the popular that entangled the relation between class and heritage as aesthetically salient analytical dimensions of (folkloric) forms.

Such transformation in the discourse of linguistic maintenance and reproduction was administered through a detailed technology of temporal management. This involved controlling the speed of transformation across historical epochs in order to control the forms of diversification: "Languages are always in a perpetual movement of transformation, such that at any period of their life that we study them, we will find them characterized by lesser or greater notable differences, although not abrupt but rather smooth and gradual ones, with respect to the previous historical period and to the one that follows" (Cuervo [1914] 1987, 23). These historical periods of language's different lives were "one barbarian or pre-classic, one literary or classic, and one critical and post-classic" (Caro [1881] 1980a, 39). These coexist in the present differentiated amongst "different social classes: the continuation, or if you wish, the posthumous life, of the barbarian usage among the rude and miserable classes that do not set foot in schools or open books; the literary splendor of the writers that are formed, as in a workshop, in the study of the best models; the critical preciseness, in the schools of erudition and philology" (39). In the first instance we have a sociohistorical temporalization of linguistic authority in the present, in which lower classes appear as temporally prior yet nominally posthumous, a mere survival, in relation to literary splendor and to the utmost authority, philology.

However, the relation between temporalization and authority is somewhat more complicated for not all languages change at the same speed. Spoken languages tend to "walk rapidly with ill–advised speed, while written tongues moderate those impulses" (Caro [1867] 1980, 359). Language change takes place more slowly among civilized peoples than among savages due to the stabilizing effects of reading and civilization (Cuervo [1914] 1987, 27). Tradition emerges here as a political temporalization of change that guarantees "gradual" and "smooth" transformations for the emergence of proper survivals. Since written words slow down the speed of change in spoken languages, then acoustic inscription through orthography becomes a technique for slowing the passage of linguistic transformation to a desirable temporality.

In order to produce this proper temporality and the appropriate word that guarantees proper descent, the philologist needs to determine the appropriate orthographic form of a spoken word. This process of intervention is guided by a different tripartite classification: "common speech is that which is used by well-educated people for daily exchange; literary speech, has as its basis common speech but appears in its artistic and, in a certain way, ideal form; and the speech of the vulgate (*el vulgo*), which we present as unformed (*grosera*) and disorderly (*chabacana*) (27–28). But vulgar speech has value precisely because it is a survival, and is potentially a link to the type of Spanish that originally arrived from Spain: "In Castilian, the vulgar speech of our days, leaving aside the arbitrariness with which it disfigures individual words, has an archaic background that represents the genuine evolution of language, free from foreign influences" (27–28). If neologisms are to be feared, and gradual language change provided by the written word provides the desired tempo of change, then barbaric survivals, once properly selected, provide ideal, archaic antecedents of words, the key to an appropriate genealogy. Such "archaic background" is the key to the link between the history of words and the aesthesis of the popular:

The popular element appears in the dictionary not only as the raw material (*materia prima*) of language, the seed (*germen*) that grows through the encouragement of literature; it is also represented by a multitude (*una muchedumbre*) of voices, metaphors, locutions and refrains that maybe have never been stamped in books. Nor could such an element be excluded: the body of the nation, the people formed the tongue, the people faithfully preserve its traditional deposit, free from the foreign, inaccessible and uncertain influences of fashion. And if at times it [the popular element] lags behind with respect to the movement of literary language, its pure and chaste speech, main current of language, feeds the private life of the wise and the literate and ties together generations regulating and assimilating the acquisitions of each epoch. That is why the creations that are proper to it and that are enfleshments of its modes of feeling and thinking, the *copla* and the refrain, have been recognized for a long time, as testimony of national use. (Cuervo 1886, xxviii)

The philologist, as the ultimate authority of language, establishes language's proper phylogenesis, an appropriate sequence of events in an evolutionary taxonomy, which ties it, as a living body, to a proper ontogenesis, an organism that grows organically, with "gradual" development. On the one hand, the speech of the lower classes needs to be corrected because proper speech (*el bien decir*) was "one of the clearest signs of cultivated and well born people" and an "indispensable condition" of all those who aspire to use, "for the benefit of others," their skills in writing and speaking (Cuervo 1886, 4). Incorrect speech was thus a sign of "vulgarity" (8) that needed to be eliminated. However, some of the words and verbal lore within that vulgar speech stood as highly valuable archaisms, monuments to the past that survived in the present and were a key to identifying the proper form of a word. Eloquence prohibits vulgarity as a spoken present yet etymology rescues it for philological use in the formulation of proper linguistic universals, accredited words valuable for all times and places where Spanish was spoken. "The law prescribes that which it prohibits and prohibits that which it prescribes" (Esposito 2009, 34). The idea of the popular, then, is born as a "defect that needs to be corrected" (32) in the sense that what constitutes it is always in need of a process of selection to guarantee its propriety. This tension of value inscribed in the popular guarantees the perpetual dynamic of renewal of the acoustic object of the popular as property and of the philologist that determines how such value is accrued.

As a science that combined a politics of descent with one of archaism, etymology turned the control of transformation of expressive culture across time into the primary link between language, different elements of popular expressive verbal culture, and literature. As stated by Caro, "etymology is not a vestige surrounded by the darkness of fabulous antiquity. It is the substantial, radical part of words (*vocablos*); that which subsists in the midst of the changes of the current of pronunciations. The etymological figure of the word is that which sustains it through different epochs and times, delaying its disappearance" ([1867] 1980, 358). The process of selection then implies discarding those changing elements in order to select only those constant "etymological figures" that persist and guarantee language's stability.

These detailed interventions in the management of orality as heritage are marked by a temporal anthropotechnology, "the governance of the prior" (Povinelli 2011). In the governance of the prior "the sociological figure of the indigenous (first or prior) person is necessary to produce the modern Western form of nation-state sovereignty even as it continually undermines this same form" (15). It is a formation that while recognizing the sociological figure of the indigenous, arrests the very possibility of its multiplicity. The identity of each person becomes the identity of the group and that of the group has to be reproduced by each person. In demanding "the supposed identity of each person with all and of all with each," it creates "a totalizing mechanism of reduction of the multitude into the one" (Esposito 2009a, 30).

However, in situations of intense creolization, the "indigenous" form that the politics of the prior attached itself to needed to be determined by a properly authorized figure, and it was not necessarily provided by the figure of the native. If Latin America and the Caribbean were the lands whose colonial histories of transculturations and creolizations continually marred original expressive purities (Trouillot 1992), then such processes demanded a zoopolitics of heritage to establish a selective process to determine proper origin. In Colombia, on the one hand, indigenous languages and peoples were submitted to a politics of the prior that reterritorialized indigenous people and languages into missionary and anthropological regimes (see chapter 3). On the other hand, the "vulgar forms" of Spanish were absorbed into the governance of the prior through a politics of descent and archaism that posed Spain, not indigenous peoples, as the originary, ideal figure. For Cuervo, Spain provided the norm for Castilian Spanish by reason of being the originary place from which the Spanish language came (Von der Walde 1997; Rodríguez-García 2010). By overcoming the zealousness of patriotic fervor that the wars of independence generated, it was possible to recognize that language united the heroes of the new republics with those of the "mother" republic (Cuervo 1935). This created the model for unity of the Hispanic American nations, with language as the touchstone of such unity.

The popular figure of speech only acquired truth-value as an archaic survival, and its truth-value was in turn guaranteed by controlling its continuation through appropriate forms of descent. Such a truth-value of popular expressions was necessary for the production of the notion of the person through a politics of vocalization but was differentially assigned to those who spoke such a language and to those who were invested in controlling its production. If one group became "the people" by virtue of speaking it, the other group became the one that governed by virtue of controlling the linguistic laws of selection and reproduction. Yet both were figures necessary for sovereignty. As such "not all peoples [and not all artistic creations] are located in the same narrative structure of social belonging, even if all people are absorbed into the same political logic of governance of the prior" (Povinelli 2011, 23).[5] Here, the relation between community and the autonomous subject is not oppositional, but rather organic (Esposito 2009). One requires the other: "we do not know how to understand the other without absorbing and incorporating them, without making them a part of ourselves" (31).

However, in a situation of creolization the problem is not only that of choosing a proper origin for descent (Spain instead of native America) or an appropriate "speed" for the "gestation" of a word. A process of selection of the appropriate form of a word amidst a history of linguistic "promiscuity," needed to be put in place. This was provided by the relationship between etymology and lexicography, inherited from the techniques of comparative grammar but adapted to the "principles" and "applications" (Cuervo 1886) of Colombian grammarians as a means "to select words/voices (*voces*) and qualify them" ([1914] 1987, 56). The method was not only used for determining the propriety of a word. It was also what led to the abstraction of ideas necessary for the formulation of the law: "All that deals with scrutinizing the laws of language is obtaining light about the laws of understanding in the generation and combination of ideas" (Caro [1867] 1980a, 373). A dictionary, then, was not simply what defined words but rather what, using the etymological methods of comparative linguistics, proposed the proper law for the usage of a term in order to enable the transcendent thoughts that led to abstractions.

The etymologist began by locating an appropriate use of vocal sources: "He who for the first time forms the dictionary or language of an uncultivated tongue (*lengua inculta*), collects from the mouth of the people who speak it all the voices he hears. From them he deduces the inflexions, classifies them, and attributes a single one that serves as its name: he then compares some sentences with others and discovers the laws of syntax" (Cuervo 1886, xxviii).

He then tracks:

> in all the manifestations of its historical development, the regulatory principles of sounds, forms, and the signification of words/voices (*voces*). He is to search for its origin [that of a word], taking into account the position it occupies in language among its cognates and with regards to those that could have been supplied by its constitutive or adventitious elements; and finally, in order to demonstrate the continuity of one word/voice (*voz*) with the source that is attributed to it, he is to know the originary languages and the circumstances that in times of transition determined the current form [of the word]. (Cuervo 1886, xxvii)

If language is a "living body," the philologist is the person who determines words' right to live. Words that are considered defective are discarded in favor of guaranteeing proper linguistic descent. Etymology then emerges as a zoopolitics of selection since "it is not possible to intervene in a positive way over the expansion and propagation of life [or of language as a living body], without having, at the same time a residual politics of death that corrects the predictable deviations of nature left to the randomness of non-planned reproduction" (Ludueña 2010, 57). The figure of patrimony is that which invokes a zoopolitics of selective reproduction of the voice by determining what is worthy of being spoken through a residual politics of death. Cuervo dedicated his life's work to developing a method for determining the propriety of words of dubious orthography when following no guide but the ear.[6] Word by word, in his dictionary, Cuervo "establishes the correct meaning of a word according to a specific context, searches for its etymology, justifies the use of each word using a great quantity of examples, analyzes it on its own or as part of a saying, annotates the variations it could have had through its use, establishes its scientific relations to other words, corrects erroneous constructions, and formulates comparisons between the respective construction in Spanish and that of other tongues" (Rojas 2004).

Etymology is a process of "decomposition and recomposition" (Sterne 2011) of the sound of words to ascertain the proper authority for transmission across a unified spoken and orthographic format. Such a process of selection and standardization provided a model for the study of other acoustic verbal expressions. As heirs to this philological tradition, folklorists in the early twentieth century copied this model of decomposition and recomposition: select a tradition's ideal form, define the boundaries of the genre through comparison of samples, and fix a proper form to represent a people through a description of its traits. Here indeed is a programmatic feature of folkloristics: traditionalists accept the change of forms, but some forms are truer than others according to a governance of the prior that provides proof of truth. The obsession with etymology in twentieth-century folkloristics in Colombia (and more generally in Latin America) is not to be confused with a rejection of processes of change in folklore. Rather what we have here is the selection of traits that establish the norm, and it is against this norm that temporal significance is established and measured.

Such a process of selection depended for its longevity on what fixed it for posterity: orthographic unity. Orthography appears as that which would bind, for the future, the process of selection of the appropriate acousticity of a word, enacted through etymology and eloquence. Caro's famous inaugural speech for the Colombian Academy of Letters in 1881, "Del uso en sus relaciones con el lenguaje" ("Of the relations between use and language"), ended with the motto of the Spanish Academy of Language: *Limpia, fija y da esplendor* (It cleanses, fixes, and gives splendor), thus making it a template for the power of *grammar* in the construction of the nation and philologists as the ultimate arbiters of the proper uses of written and spoken language (Von der Walde 1997; Rodríguez-García 2004, 2010). If philologists were the custodians of speech and speech was one of the primary arbiters of national unity by signaling the presence of "a" people, then the most appropriate sovereign was the philologist who identified how those people should speak and write. Philologists (scientists, not literary figures) were able to differentiate between "customs" and "science," between "use of language" and "abuse of language" (Caro [1881] 1980a, 16). Even though this was an inaugural speech for the foundation of the Academy of Language, its role was to institutionalize what was, by then, a hierarchical division that was of considerable concern to differentially placed lettered persons.

In 1835 Juan José Nieto, a mulatto (or pardo) member of the Cartagena Caribbean political elite, complained bitterly to President Francisco de Paula Santander of how Andean "authorities now 'vomited their rancor' against the port city and 'shockingly ridiculed cartageneros for their alleged illiteracy,' their 'way of talking' and 'their customs'" (cited by Helg 2004, 237). A sign of how the growing rivalry between the Colombian Caribbean and the Andean region, even at such an early date, was being named through a politics of the voice that left no uncertainty about the increasing importance of the notion of culture, understood as a property of the person expressed through the acoustic parameters of the voice.

Let us then recast the definition of orality posited at the initial part of this chapter. Orality is that historical mode of audibility that, through the anthropotechnologies of philology—eloquence, etymology and orthography— turns vocality into property through a politics of the governance of the prior and a zoopolitics of selection. If orality is centrally tied to an idea of the people and to an aesthetics of how they should vocalize, then such a notion of the people is, in turn, tied to the property that expresses their belonging. Folklore becomes a thing to be collected and to produce identity after being properly identified by corresponding authorities. Culture is what needs to be taught to the people to guarantee an appropriate person but is also what comes from the people, with the authority of the grammarian. To determine that heritage: the interactions between eloquence and etymology produce mutually constitutive figures of the notion of culture. One immunizes the other. If eloquence prevents an improper use of speech, etymology prevents an improper ascription of heritage. These different notions of culture do not oppose each other; rather they are "superimposed one on the other" (Esposito 2009, 31). Through the standardization of orthography the relation between them is inscribed into property. The place of orality in community thus ceases to be understood as a type of obligation and becomes a propertied exchange, so crucial for the production of a modern productive rationality (Esposito 2009). While this is a zoopolitical history of distinction and selection through a genealogical reorganization of hearing, it is also a history of media through the creation of a voice that embodies property. Genres of literature or folklore understood as objects or "works" could only exist by virtue of this mediatic process.

A uniform orthography was the precondition for the transition from the vocal to the literary or the folkloric as a form of the literary. It demanded the generation of a standard as the rational means for the expansion of the medium (Sterne 2011; Del Valle 2002). The creation of standards is a central aspect of the rise of regulatory development necessary both for national policy and industrial protocol. It transforms sensorial acoustic dimensions into "works" or "things" that can be disseminated through a uniform format that also allows them to be inscribed in a transnational mode of labor and production (Sterne 2011). In the available technologies of inscription of the period, the ideal of orthography as a medium that provided unity for the dissemination of acoustic intentions became a model of dissemination for other acoustic art forms, particularly music. The understanding of orthography as a technology of inscription that allowed for the proper dissemination of voice, and therefore its transformation from speech to literature, was so strong in late nineteenth-century Colombia that it was also used as a model for both vocal and instrumental music.

ORTHOGRAPHY AS MUSICAL NOTATION

Diego Fallón (1834–1904), a recognized poet, composer of salon music, and bandola player, published two books on his own system of musical notation based on orthography: *Nuevo sistema de escritura musical* (New System of Musical Notation) (1869) and *Arte de leer, escribir y dictar música, Sistema Alfabético* (Art of Reading, Writing and Dictating Music, Alphabetic System) (1885). In them he developed a system of musical notation based on alphabetic writing that radicalized the relationship between acousticity and orthographic inscription. He tried to translate every aspect of musical sound into orthographic notation in order to propose it as the viable form of musical dissemination in the nation. He thought the lack of a proper infrastructure for music production in Colombia could be addressed by turning staff notation to an orthographic one, easily adapted to typographical machines (Fallón 1885, 2). This notation system was meant as a new format that supposedly facilitated the wider diffusion of music through the newspaper. He used the newspaper chronicle, which provided the model for professionalization of the writer (Ramos [1989] 2003), as the model for professionalization of the musician.

The first musical scores published in the second half of the nineteenth century appeared in newspapers in cities such as Bogotá and Medellín (Bermúdez 2000; Velásquez Ospina 2012), making it a site of musical professionalization through technological inscription. But staff notation was difficult to print with typographic technology and its use was limited. Fallón expected orthographic notation would do for music what it was doing for literature:

> One is amazed at the profusion with which different types of newspapers circulate, be they scientific, political or of news, and the amount of literary productions of all kinds found in even the most remote villages of the Republic. But one notes that, on the contrary, musical works and methods destined to the cultivation of this art lack the expedite, fecund and economic medium of publication that is felicitous for politics, literature and science. . . . This fact has been and is today so gravely deterring to the interests of musical art, that one only needs a little bit of common sense to deduce the urgent necessity that a country with a small popu-

lation and few resources adopt a new procedure for the publication of music. (Fallón 1885, 9)

In his two books Fallón linked a new mediatic imagination to the idea of the musical work, to a method of teaching it, and to linguistic inscription as a model of dissemination. In the process of designing his notation system, he generated his own concept of the musical work. Contrary to the development in Europe where the emergence of the work concept in music implied its emancipation from language (Goehr 1992), here the emergence of the work concept implied recasting the conservative relationship between music and language as a mediatic, modern one. Such a relation was understood as one of "translation," (a word that appeared in his musical scores, see figure 4.1) between staff notation and orthographic musical notation.

Let us recall that orthography was understood as a process of uniformization that translated multiple pronunciations into a single orthographic sign. Likewise, the two modes of musical notation could be understood as two formats that could be translated one into the other to produce an appropriate mediality. For Jonathan Sterne, mediality evokes "a quality of or pertaining to media and the complex ways in which communication technologies refer to one another in form or content" (2012, 9). The term indicates "a general web of reference" that is essential to understanding how expressive forms "represent, figure and organize broad realities and relationships" and to "a collectively embodied process of cross-reference" in the arts (2012, 9–10). A format, on the other hand, "denotes a whole range of decisions that affect the look, feel, experience, and workings of a medium. It also names a set of rules according to which a technology can operate" (7). In writing, the format is the particular form of inscription—not only the type of alphabet chosen, but what the signs chosen are meant to represent and how they are meant to be used. The relationship between orthography and the newspaper was that of a format and a medium. Fallón operated a translation between different written formats—from staff notation to orthographic notation—in order to purportedly transform the musical work's mediality and adapt it to the technological infrastructure available in Colombia. His method generated a radical auralization of orthography through its musicalization.

But he was not the only one generating ideas about musical aurality through links to orality. The turn to the aural in the nineteenth century implied an acoustic imagination of the body modeled on the different apparati that were being invented through experimentation with listening—from ear trumpets to stethoscopes—and vice versa (Sterne 2003; Steege 2012). Both Cuervo and

Caro repeatedly compared the physiology of vocal production to a musical instrument: "The apparatus with which the human voice is produced is a musical instrument of great perfection. The lungs, aided by the diaphragm, drive the air as they would a bellows; the larynx functions as a sonorous tube; the mouth and nasal cavities strengthen and modify the sound" (Cuervo [1867] 1987, 103). Or as stated by Caro, the "apparatus" that produces the human voice "is constituted by a great cavity that has diverse organs which are the interior and superior palate, the tongue, the teeth and the lips. This apparatus is a marvelous instrument, and speech (*el habla*) is a species of song or vocal music" (*una especie de canto o música vocal*) (Caro [1928] 1980, 331–32). Here, the organs of vocal production are mechanically understood. The aesthetization of the "physiologized human" that emerged with the experimentalization of hearing in the late nineteenth century (Steege 2012) also implied a mechanization of physiology. The understanding of the voice as a machine that could be repaired in case of failure (Connor 2006) or made to run smoothly through appropriate use, complemented eloquence's mechanistic understanding of pedagogy as a transformation of the body's vocal disposition through rote repetition.

The figuration of music into orthography by Fallón was not understood solely "as the simple substitution of signs used for musical notes for the characters of the common press (*la simple substitución de los signos de la nota por los caracteres de la imprenta común*) (Fallón 1885, 10). The system was seen as containing "innovations, simplifications and practical results" (10). The second book *Arte de leer, escribir y dictar música, Sistema Alfabético* is divided in a series of lessons that through a numbered question and answer format teach the fundamentals of music theory along with the new system of musical notation. Like the philologists' proposals for eloquence and orthography, the relation between pedagogy and circulation yields a process of regimentation of the relationship between the voice and language now also linked to music performance. What emerges is a new understanding of the *musical* ear that is modeled on language, an aesthetization of the mechanistic understanding of the physiology of the voice. It is impossible to fully explain Fallón's system here. But understanding some of its principles helps us explore how this musicalization of the relationship between voice and hearing is created lesson by lesson, extending the mechanistic understanding of physiology to the mechanics of pedagogical regimentation of instrumental (not only vocal) musical works.

Fallón's method is based on a theoretical distinction between sound, time, and value as the principal elements of music. Value is "the numeric relation that each sound or chord, each dotted note or silence or pause in a musical pas-

sage, maintains with the unit of time that has been agreed upon in order to provide the measure or form of comparison" (1885, 13–14). Sound is not defined,
but its conception appears in the way it is notated. "Sound" is represented by
consonants, silence by the h (also silent in Spanish language orthography),
and time value (duration) is represented by vowels. In addition, each alteration of pitch (flats and sharps) is assigned a specific letter. Thus, the chromatic
scale, beginning with the note C, is represented by the letters B D F G Y L Ch
N V R S T.[7] In lesson 2 we find this initial distribution of sounds into letters:

13. Q: Which consonants represent the five altered notes? (las cinco
notas accidentales?)

A: The following in ascending order, that is from the lower to
the higher:
D G Ch V S

14. Q: Which consonants represent the other seven notes called
natural in the ancient system?

A: In the same order, they are the following:
B F Y L N R T

15. Q: Recite in a progressive series all the notes of the new system
from the lowest to the highest:

B D F G Y L Ch N V R S T. (1885, 15)

The first thing we notice is that each letter is assigned to a specific pitch, regardless of its function within the tonal system. The model is mechanical,
like that of a musical instrument. This system was not meant to represent the
relationship between notes in tonal harmony. For example, lesson 3, questions 18 and 19, the student asks, how many ways can C# be written in the
ancient notation? The professor answers, "in six ways." Fallón outlines what
he considers all the enharmonic notations for the pitch and simplifies them
by suggesting that in the orthographic notation, all the superfluous profusion
of symbols to express one single "sound" or pitch be eliminated by a simple
substitution for the letter B. For Fallón, the history of the rise of tonality lay
elsewhere and was not important in the circulation of music.

The implication of this dismissal of tonal structure for understanding the
relation between words and music is fundamental. But let me first finish explaining other dimensions of this notational system.

The different durations or values are represented by a vowel or combinations of vowels:

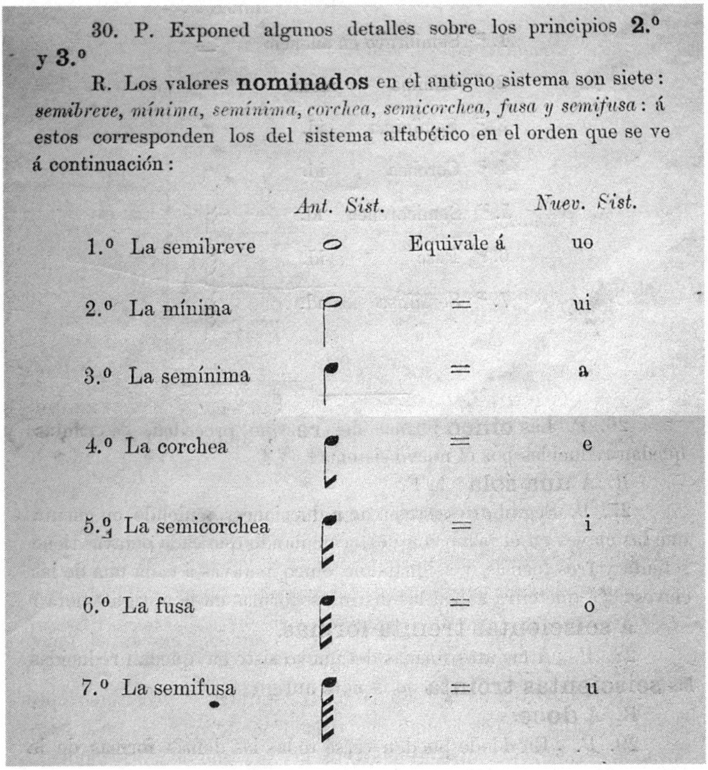

30. P. Exponed algunos detalles sobre los principios **2.**°
y **3.**°

R. Los valores **nominados** en el antiguo sistema son siete:
semibreve, mínima, semínima, corchea, semicorchea, fusa y semifusa: á
estos corresponden los del sistema alfabético en el orden que se ve
á continuación:

		Ant. Sist.		Nuev. Sist.
1.° La semibreve	○	Equivale á		uo
2.° La mínima		=		ui
3.° La semínima		=		a
4.° La corchea		=		e
5.° La semicorchea		=		i
6.° La fusa		=		o
7.° La semifusa		=		u

FIGURE 4.1. ▪ Diego Fallón. 1885. *Arte de leer, escribir y dictar música, Sistema Alfabético*. Bogotá: Imprenta Musical. Courtesy of Centro de Documentación Musical, Biblioteca Nacional de Colombia, Ministerio de Cultura.

In notation, the vowel that corresponds to the value of the note usually follows the consonant that corresponds to the pitch (which he calls sound). Silences and dotted values are also assigned a letter and different octaves are indicated by print type, covering a total of five octaves: italics and lowercase letters for the lowest octave of the piano, regular type and uppercase letters for the next, and so forth. Since all notes have both pitch and duration, the notation of each note will end up being a syllable: a consonant that represents the pitch plus a vowel that represents the duration.

If we look at a simple score, for example, we are able to see some of the correspondences with the above explanation. Just to take the first bar of the left hand, c, e, g, e in eighth notes corresponds to be ye ne ye: the letter that gives the pitch plus the letter that gives the duration. This transformation is graphically depicted in this example:

FIGURE 4.2. • Diego Fallón. 1885. *Arte de leer, escribir y dictar música, Sistema Alfabético*. Bogotá: Imprenta Musical. Courtesy of Centro de Documentación Musical, Biblioteca Nacional de Colombia, Ministerio de Cultura.

The notation of chords (or simultaneous musical sounds in general) is modeled on that of words:

74. Q: Before illustrating how melodies with accompaniment are written, tell me, how should chords be written?

A: In the same way that we write any word, taking into account that the lowest note of the chord is to be the first letter of the word.

75. Q: After writing this first letter in what direction should one go?

A: In horizontal direction to the right, as in melodies.

76. Q: If that is so, then how does one distinguish between a chord and a melody?

A: In that the chord always finishes in a consonant, its letters are always tied together as in a word, and they contain more consonants than vowels and some of these consonants can be united without an intermediary vowel.

77. Q: What should be the second consonant of the word that expresses a chord?

A: That which follows the lowest one of that chord, moving
from the lowest to the highest tone.

78. Q: Which the third?

A: For the third, fourth, etc. we proceed in the same order, from
the lowest to the highest such that the last consonant of the
chord represents its highest note. (Fallón 1885, 33)

The initial consonant that represents the lower note of the chord is followed
by the vowel that represents its duration (or "value" according to Fallón's no-
menclature). That initial consonant also has the character type (i.e., uppercase, lowercase, etc.) that corresponds to its octave. This determines the duration and exact altitude for the chord. The other notes of the chord are simply
consonants expressing pitch, but the vowel that corresponds to the value of
the note is repeated after the second note/consonant of the chord in order to
"facilitate its pronunciation" (1885, 34). Chords and words are understood as
significant units of sounds whose unity is reflected on the page orthographi-
cally. For example, a C major chord with a duration of a quarter note and
with an altitude corresponding to the central C of the piano becomes *Bayan*.
B gives the pitch and the octave, *a* gives the duration, *y* corresponds to E, followed by an *a* that repeats the duration, not because it is needed but in order
to create a syllable that can be pronounced, and *n* represents G. Bayan equals
C major.

In translating every parameter of musical sound into orthography, Fallón
created an analogy between different aspects of music and different aspects of
language: consonants, vowels, typographic elements of alphabetic notation,
words, and more. Music is "linguicized" and language is musicalized. The ultimate translation occurs not in writing but in the pedagogical method. The
resultant syllabic scores were meant not only to be read but also to be recited,
a gibberish of syllables corresponding to a particular musical piece. Supposedly that process aided in memorizing the piece and performing it through
a system analogous to Marroquín's spelling *coplas*. It is as if Fallón wanted
to divest musical notation from its subordination to sight by translating its
effectiveness into a necessary pedagogical passage through another acoustic
language. Although all of Fallón's examples come either from Western classical music or from European-derived salon dances that were then in vogue
in Bogotá (pasillos, polkas, etc.), the genealogy of philology weighed more

heavily on him than the genealogy of Western tonality, thus language rather than music provided his canon for musical notation and pedagogy.

One has the sensation that what is happening in this translation process is a venetriloquization of music by having it speak in another medium's voice. This is seen especially in the value given to syllabic recitation as a mode of learning an instrumental musical piece. Music acquires presence through transference to another *corps sonore*. Music becomes a "mixed body" and the voice "an organ of listening as well as transmission, impression as well as expression" (Connor 2006). For Fallón, in fact, such a mechanistic pedagogy of musical syllabics was supposed to bypass some of the perceptual difficulties in learning music. A final example, to illustrate this mixed body of the musical, this time from lesson 62:

186.   Q: What is a measure?

A: In music, a measure is employed in various ways. So, in order to make the definition of the term understandable, in each of its definitions, we need to begin with a practical understanding of the matter.

187.   Q: In pronouncing the word "solo" what fact do we verify in terms of musical time?

A: That the interval [of time] that passes between the *o* of the first syllable and the *o* of the second, is equivalent to the duration of value that in music is called sixteenth note (semicorchea).

188.   Q: Can we say the same word only in thoughts? (Se puede decir la misma palabra con solo el pensamiento?).

A: Yes, sir.

189.   Q: Can we then measure sixteenth notes only with our thoughts?

A: Yes, sir.

190.   Q: Can a person who has no musical ear (*que no tenga oído para la música*) pronounce the word *solo*?

A: People who have musical faculties as well as those who lack them know how to measure one of the musical values with precision, as long as they know how to speak.

191.   Q: Can the person who knows the Credo recite it mentally?

A: Yes, sir.

192. Q: According to this, a person who has no musical ear can measure sixteenth notes through his thoughts?

A: Yes, sir. (1885, 105–6)

Fallón's syllabic notation was supposed to facilitate musical perception simply because humans possessed the faculty of speech—a training not only of the
physical ear but also of the mental one. The syllabic method for music theory
intervenes in "the malleability of personal identity" (Schmidt 2000, 136) by
mediating between the mechanics of articulation of musical time and the inner voices we hear. Not only were sounds turned into manipulable syllables
by their inscription on the page and by the voice. That very act of translation
(Rodríguez-García 2004, 2010) was supposed to provoke an awareness of
the enlightened acoustic capacities residing in people's innermost thoughts.
Through the ear in the mind, music became the art of transcendence precisely
because of how easily a mechanical gesture of the voice could give access,
through acoustic analogy, to our unacknowledged musical natural dispositions.
The idea of an "inner voice" of the mind provided the template for musical
understanding by making the acoustic analogy explicit through rote syllabic
recitation of musical sounds.

So let us now return to the question of the relation between music, words,
a mechanistic understanding of the body, and the notion of musical work that
emerges here, and that for Fallón was crucial for music to acquire the same
import as literature in the constitution of the nation. The emergence of the
idea of a musical work in Europe in the eighteenth century was partially tied
to relating the musical form to the autonomous individual. The idea of transcendence implied turning music into the most autonomous of the art forms
in two senses: it was that which contained all the mechanical laws of its rationality within itself—music understood as the most autonomous of artistic
objects since it did not "represent"—but it was also that capable of reflecting
the innermost self, of embodying the subject's very ontological nature. Music
was no longer to be understood as a language that magically and mimetically
mirrored the structure of the cosmos and of humans, but rather as a science
that contained all the laws of signification within the laws of its own system.
As such, "instead of being interlaced with things music became a thing itself"
(Chua 1999, 78). As an empty sign, music came to be validated in new ways,
its formal laws as a psychoacoustical mechanics of the inner subject, and instrumental music's incapacity to represent became the ideal for the rational
autonomous subject. But if the body of music—its mathematical form, especially of instrumental music—reflected its mechanical, rational structure,

it was the idea of voice, that acoustic element that moved between the world and the body, that came to represent musical passions, the capacity of music to "move" the soul:

> It was vital that reason controlled music by an act of naming lest music should return as some animistic spirit that the mind had supposedly expelled from the body and stir up the passions beyond the limits of reason. The mind as the agent of the soul disciplined the musical motions within the body, subjecting the movement of the passions to the precision of the concept. Words were therefore as much a sign of moral strength as of epistemological truth; they instrumentalised desire. This question of morality connects the realm of innate ideas to a second area of mediation, that of the passions. If the body was to have any significance for the thinking ego then it would be at the point of contact with the soul, and it was at this meeting place that music found itself embroiled in the moral physiology of the passions. (Chua 1999, 85–86)

Music came to reside "in that precarious site of the passion that mediates between the activity of the soul and the passivity of the body" (86) and the emotions were understood as having an "isomorphic connection" with the body (86). In such a mechanical arrangement "it did not matter whether one was a Cartesian moralist or a material sensualist, the effect of sound on the body was the same" (86). Thus, "the moral dialectic is therefore a tension between the material sensations and the power of rational control" (86). Vocal music, "as the passionate script of the soul, provided the data for the linguistic and moral origin of humanity" (88). Music was an imitative art because it expressed the soul's pure expressivity and the voice was the model for this. The theory of Baroque musical affects was "not only a symptom of an age of representation but a matter of body control" (87). Music becomes a "wordless rhetoric" of expressivity, and instrumental music came to embody that rhetorical ideal.[8]

By the nineteenth century such a mechanical relation of sounds to the body was transformed by the rise of the understanding of the body, and especially the ear, as a physiological, biological instrument with the capacity of sensation and perception. It mediated the relation between the body and the soul through a psychophysiological understanding of the relation between mind, perception, sensation, emotion, and body.[9] Thus psychoacoustics enacted a movement from the voice as a model for music's effects on the human to the ear as that which channeled all its impressions (Sterne 2003; Steege 2012). This "refunctionalization of the ear" made it "an instrument and a field of observation" . . . in which the listening person "is both subject and object . . . capable of making sonic perceptions present to itself and also constrained to gather information about the acoustic world indirectly, by obliquely observing itself in the act of listening" (Steege 2012, 44). This was a crucial move, as Steege explains: "reimagining the ear as a multiple and malleable medium, whose design would simultaneously limit and expand the possibilities available to aural experience, was perhaps the most consequential change Helmholtz introduced into discourse about hearing—animating it as a human function while rendering it vivid as an object of knowledge. To think of the ear as at once instrumental and educable: this is the often overlooked problematic that largely defines the Helmholtzian project and its specific aurality as unique" (Steege 2012, 44–45).

It was not by chance that Caro cited Helmholtz's *On the Sensations of Tone* in a footnote in his "Manual of Elocution" (Caro [1881] 1980, 447), the most musically imbued of his philological writings. Because, as we have seen, for Caro as for Fallón, the ear was also instrumental and educable. The epochal transformation of the relation between music and listening that takes place between the seventeenth and nineteenth centuries then, is not solely articulated in Germany. What we see emerge is an ontological relation between a physiological understanding of the body (its "animal vitality"), an instrumental relation between the ear and the training of the body, and the turn to music as a moral force to define the inner relation between the rational, the moral, and sensations in the autonomous subject. For Lévi-Strauss it is the historical moment when music becomes Western cosmology's own myth by becoming the embodiment of the transcendental autonomous subject, in whose innermost being reside the struggles between sensation and rationality, between force of feeling and scientific knowledge, and music as the art form that mediates between both (1985). So let us return to the implications of this for understanding the different relation between language and music through which this takes place in Colombia (and arguably in Latin America).

Fallón had his method approved by the National University as the desirable one for teaching music and managed to publish a series of scores in it. The 1885 book is prefaced by Narciso González Lineros, senator of the republic, director of several newspapers and a primary figure in the organization of elementary education in the state of Cundinamarca (next to Bogotá). It is also given full support by the Consejo Académico de la Universidad Nacional (the directing academic body of the main public university), which states that the method of writing should be taught in schools and to that effect stipulates that "a work should be published in the form of a text for teaching that explains the new system [orthographic notation] through the antique one [staff notation] and that presents all the examples destined to explain the rules of the new system" (Fallón 1885, 10). This is tied not only to a politics of education but, as González Lineros makes evident in the preface, to an ethical ascription to music that is quite different from that of the literary:

> Print has transformed the world making modern societies more instructed, richer and freer than ancient ones. But it has been able to do very little in terms of polishing and softening the character of men, making them benevolent, charitable, truly sociable and has been impotent to create shame around vice and extirpate it. . . . the fine arts, and in particular music are, without doubt, those called to soften human character and to serve as a dam against the progresses of alcoholism. Take the cultivation of music everywhere, and the tavern will be left abandoned, at the same time that the home will triumph over the club, because it is known that music is the most powerful element of sociability; and the egotistical passions fomented by the incomplete civilization of which the nineteenth century boasts, will give way to the sweet, generous, compassionate affects awoken in the heart by the cultivation of the divine art. (Lineros, in Fallón 1885, 5–6)

If the literary was what gave distinction through correct speech, it was music, "the most powerful element of sociability," an art understood as able to awaken "sweet, generous and compassionate affects in the human heart," that would be the standard bearer for the ethical training that would lead to the completion of the "incomplete civilization of which the nineteenth century prides itself." If language was to generate distinction and guarantee proper inheritance, then music was what would provide the means to temper character. For Lacoue-Labarthe and Nancy (1988), the literary is what provides the site for critical thinking. But music as an art form would partially remain outside such critical scrutiny by ascribing musical culture to the "conventionalizing processes of morality" (Wagner [1975] 1981, 42). Music and language acquire a curious alignment in this modernization of their relation. On the one hand, music acts to temper the uncontrolled passionate dimensions of the self because of its capacity to affect a vital body, but the linguistic pedagogy of the musical provides the instrumental tools for establishing the relation between the inner mind and the instrumental training of the ear in order to figure an appropriate moral subject.

As we have seen, the nineteenth-century postcolonial elite was obsessed with organizing nature through botanical and geographic expeditions that made it available for classification and political and commercial control (Nieto Olarte 2007). This generated a vitalist understanding of nature, partially in the postcolonial exchange between different forms of enlightened vitalism between the Americas and Europe (see chapter 1). If the body was central to the expression of animal elements in the human, as we saw stated by Caro at the beginning of this chapter, then the voice needed to be regulated in order for that human to become a proper person, and the passions needed to be tempered in the moral alignment of the physiological body. Adapting to our context Chua's analysis of the regulation of affect tied to the voice, we can see that the mechanical dimensions of a Baroque understanding of the relations between passions, reason, and specific sounds is also present here but it takes a different form. The dubious knowledge of the ear and the variable acoustic gestures of the voice are trained by music in order to make them instrumental to the moral rendering of passions in a vitalist understanding of the body. The rationality and expressivity of the eloquent voice and the sweet affect of divine music intoned the complex networking of affect, reason, and expressivity as mutually constituting relations in a Baroque vitalism that entwined the musical and the literary. As such, in Latin America, the emergence of the idea of the musical work did not imply overcoming the association of music and language through the voice, but rather their careful alignment in a Baroque, vitalist conception of the acoustic ear that could be mechanically trained through the anthropotechnologies of the voice and of writing.

The idea of cultural policy developed a notion of citizenship as one based on a subject in need of ethical training, a central idea of governmentality (Miller 1993). By building on the historical analogy of music as a form of language Fallón provided not only a mode of inscription in the physical act of copying/translating notes but also a technology for training a person's moral dispositions. Voice/music becomes the site of enacting a proper moral relation between different dimensions of the acoustic. Just as voice can be understood as the representation of a person's distinction, music can be understood as that which provides the proper moral comportment for the person and for the people. The proliferation of a music genre that is understood as immoral could be understood as a sign of a people's social moral decadence. The control of the *corps sonore* of the social group thus was also a central aspect of its healthy immunization.

Fallón's writing method can easily be relegated to the fantastic inventions of the absurd, especially in its aspirations to a universalism that would sweep away staff notation, a technology of replacement of what he saw as "an antique" and highly problematic mode of musical notation. The history of "new" media is filled with failed technologies (Gitelman 2006), inventions that at the moment of their creation seemed to fulfill wants and needs and were marked by the intensity of the new and by the affect of discovery. Technologies that rather than fulfilling their destiny disappeared into the dustbins of ephemeral contraptions. But the political and cultural significance of what they sought to put into place extends way beyond them.

In the nineteenth-century writings on vocal anthropotechnologies explored in this chapter, the voice emerges, again and again, as an ambiguous element. If eloquence, etymology, and orthography continually aim at creating an expedient voice with clear performatives that represent a particular zoopolitics of the person and of the people, voice always appears simultaneously as that which potentially disrupts such expediency. A gap emerges here between the recognition of the voice as a powerful emanation and its training into propriety.

Part of what defines the human as a species is the capacity to transform the sounds they emit into multiple possibilities of relationality and signification between different entities. That is why an immunitary protection from voice's power is common to several metaphysical traditions, and also appears as a central element of magical power and of the shaman's capacity to act (Napier 2003). The injunction to protect oneself through prevention is not necessarily negative. What is described in the history of this chapter is rather a type of immunity, historically and polemically associated with Western modernity, which creates a problematic political distinction between the human as a species and the juridical person such that only those people who fit the criteria of juridically defined person have a worthy life (Esposito 2008). A specific history of metaphysics in the West understands governmentality as a transformation of the "animal" nature of the human into full-fledged personhood (Sahlins 2008). In such a zoopolitics, the real site of orality is not necessarily "the other" but rather an understanding of the grammaticalization of the voice as a potentially failed prescription that requires a permanent process of immunization against its very power. Voice's containment moves between the pragmatism of the use of its force through pedagogy and the phantasmatic infrasound of its expressive potential going awry. The different aspects that constitute the "oral" always name not only a specific anthropotechnology (a pedagogy of pronunciation, a control of linguistic descent via etymology, and so on) but also a spectral dimension that is beyond full apprehension and containment through its immunized deployment. This disjuncture between anthropotechnology of the voice and its spectral figures became precisely the place of theorization of orality through its repeated occurrence in scholarly discourse. More than signaling a different mode of speech and a technique for creating social inequality, more than an epistemic mechanism of purification meant to create social and political distinction in a particular place and to mark as different (or "other") a specific people (Bauman and Briggs 2003), the place of orality appears rather as that recurring scholarly discourse about the disjuncture between an immunizing inscription and the materiality of the voice, generating an ontology of the voice as a failed writing, a failed prescription. This is its ultimate epistemic and ontological site for nineteenth-century Western metaphysics. By signaling this, I do not wish to erase the use of orality to create social inequality in different moments of history and in different places. Rather, what I wish to point out is that it is always readily available to be recurrently cast as a disjuncture or as a failed prescription precisely because its site of articulation, even today, more than a century after the invention of recording machines. The voice then played a crucial role in defining the idea of culture simultaneously as a form of recognition of the other and as a form of assistantship needed in order to transform the other's failure into the proper person. Such a melancholic immunization of folklore actually made it impossible for it to exist—it is always on the verge of disappearing in order to be rescued once more. As one of the forms of authority for the prescription of an appropriate acousticity of the ear in the voice, the intertwined ideas of culture that emerged from such a zoopolitics have had a long-lasting influence in Latin America and the Caribbean understanding of the popular.

THE ORAL IN THE AURAL

In his "Manual de Elocución" (Elocution Manual), Caro, the modern conservative Hispanist, seems to predict the historical movement from the emphasis on the oral to the emergence of the importance of the aural that is currently taking place with the consolidation of the field of sound studies: "the recent inventions of the phonograph and the telephone have given rise to very interesting descriptive works on acoustic phenomena" ([1881] 1980, 446). The contemporary intensification of the aural can be seen in the gradual institutionalization of the field of sound studies in the North (Sterne 2012a) and the rise of diverse types of graduate programs of music/sound studies in Latin America and the Caribbean. This is deemed part of an "aural turn" that acknowledges the increased presence of sound as field of theorization. This book took form in the midst of this epistemic transformation. But what questions does the history of the voice, so frequently reduced to an ellipsis between orality and aurality inscribed on the page as orality/aurality pose for the way such a turn is being described? To begin to address this question it is perhaps good to consider what Caro's reference to sound technology implies. As part of the above passage, Caro also states: "If we study the way the sounds of language are formed by the vocal organ, as any other sounds are formed through the medium of the instruments that produce them, followed by the propagation through the air or other elastic body; the physical conditions that determine intensity, pitch, and timbre of those sounds; the form of the vibrations that conduce them, examined through optical means, this study of phonetics is scientific and pertains to the physical sciences" ([1881] 1980, 446).

Caro elaborates on the nature of "descriptive works regarding acoustic phenomena" by distinguishing the functions of science and art with regard to vocal sound. Besides phonetics, which belongs to the physical sciences, there is physiology, which addresses the study of the organs that produce the voice. The latter, as a science, also deals with the ear, which seems to lie ambiguously between physiology and physics because it is an organ of "perception of sounds, transmitted in the manner taught to us by physics" ([1881] 1980, 447), a sentence that he footnotes with Helmholtz's *Théorie physiologique de la musique*, originally published in 1868. But, says Caro, even though such scientific knowledge is "antecedent and a luminous auxiliary to the art of elocution" (447), it ultimately does not contribute to its practice because "the exercise of an art does not require or necessarily presuppose the possession of the respective science" (447).

Caro addresses the need for multiple disciplinary agendas in the study and practice of sounds (especially of the sounds of the voice) even though he calls for their functional separation. While the physical and physiological sciences explain the acoustic dimensions of the voice and the ear, elocution forms part of the arts, a realm of "practices" with sounds that, as in other passages of his work, metaphorically aligns the production of the voice with the mechanics of the production of sound in musical instruments. Yet, as we have seen throughout this book, keeping separate the scientific and artistic aspects of sound, or, to frame it in other terms, those aspects related to nature and those pertaining to culture, requires intense philosophical and political labor.

Caro's easy slippage from studies of phonetics to questions of acoustics shows us how easily in the late nineteenth century, questions of the physics of sound and, ultimately, of ontology of the voice, became epistemological questions about how to parcel out the attributes of sound to different disciplines. Moreover, since the purpose of the "Manual of Elocution" is to train people into *bien decir* as a politics of proper citizenship, the above quote also implies a particular politics of implementation of artistic practices. If Caro's writings, as we have seen, have to do with the mediation of language into the political and juridical sphere of the nation-state, his modernizing spirit led him to posit, even then, a potential transformation of the oral into the aural. If his work is foundational to the emergence, in late nineteenth-century Colombia, of a zoopolitics of the voice deeply imbued in the theologico-political, his nod to the aural through the rise of new technologies and through the separation of the sciences and the arts, simultaneously invokes the use of the voice for the techno-developmentalist and rational-scientific division between the sciences and the humanities so central to the aura of modernization.

What is interesting about the initial quote in this chapter is that it shows us how the history of the oral and of the aural are imbricated in each other. As stated by Jonathan Sterne, "we live in a world whose sonic texture is constantly transforming and has been for centuries" . . . and "as sonic worlds have changed so have the conceptual infrastructures writers have built to behold them" (2012a, 1). The field of sound studies has been seen as an "'emerging field' for the last hundred years" (Hilmes 2005, 249). And yet the difference between the moment when Michelle Hilmes wrote and today (2013) is that, despite its broad transdisciplinary base, sound studies are already taught in institutionalized programs of study and are widely recognized both in the North and in the South as a relatively consistent, even if emergent and dispersed, disciplinary terrain with bibliography, blogs, meetings, publications, artistic events, and other types of formal and informal activities that coalesce around its rubric (Sterne 2012a). Evidently, in the midst of a long, continued emergence, something is changing, a change indicated by the intrusion of the name of the discipline, sound studies, even with its own *Sound Studies Reader* (2012a). To give a name to a field is to recognize the power it has "to make us think and imagine" but it also makes one wonder how to characterize the type of protagonist that has entered the scene (Stengers 2009). Becoming aware of the historiography of a field is a means to characterize it, "not in order to deduce the present from the past but to give density to the present" (35).

This book originally began as an inquiry into the politics of knowledge about popular music in Latin America and the Caribbean. It became a book about how the relationship between practices of listening and vocalization were central not only to configurations of knowledge about music or language but also to the politics of differentiation between the human and nonhuman. In the case of humans, the voice is materially constituted simultaneously through the body (by means of vibrating vocal chords) and the world (by means of air that makes the chords move) yet does not fully belong to either. As we have seen throughout this book, this ambiguous materiality and ambivalent ontological belonging places the voice as a phenomenon that hovers at the juncture of the differentiation between the human and nonhuman and that mediates between the world and the person, subject to both techniques of power and techniques of the self. Because of its dispersed materiality and ontological ambivalence it has the potential to highlight the limit of the effects of the technologies of power at the very moment they are enacted. This is, of course, not the only object, phenomenon, or force to do so. As Bruno Latour showed us, the disciplinary process of constituting objects as "separate" is a central history of modernity that reveals how, at the very moment of disciplining or purifying a particular sphere, the limits of such separability also emerge, a process that led him to ironically propose that "we have never been modern" (1993). But some objects or forces are less amenable to the invisibilizing processes of disciplinization and more readily reveal modernity's conundrums. What is particular about the voice is that, as a force that hovers between the world and what humans do with the world, it is particularly poised to be used as a disciplining force and yet it simultaneously easily reveals the limits of such a process. It is precisely because of its material and ontological ambivalence that the relationship between the ear and the voice appears just as frequently associated with the history of techniques of disciplining and purification so central to a Western metaphysics of transcendence and to cosmological histories that understand such a history otherwise (Connor 2000; Schmidt 2000; Hirschkind 2006).

In this book I have used Fabián Ludueña's term anthropotechnologies to depict the different processes of constitution of the voice as an instrument of political transformation of the boundary between humans and nonhumans in the history of hominization. This term brings to the foreground the biopolitical slant of Foucault's idea of governmentality by emphasizing the historical process of the constitution of the human species as a product of the techniques through which the "species *Homo sapiens* acts over its own animal nature" (Ludueña 2010, 11). Since the relationship between the ear and the voice mediates many areas that are central to the political definition of life, it appears again and again as defining the zoopolitical. As different histories of the voice and of the ear accumulate, it becomes increasingly evident that this acoustic zoopolitics has been a crucial dimension of late eighteenth-century and nineteenth-century modernity (Sterne 2003; Erlmann 2010; Steege 2012). Jonathan Sterne gave the name of "audile techniques" to the use of listening as a learned tool and a "skill to be used for instrumental ends" (2003, 93) such as listening for medical symptoms through the stethoscope or for specific musical sounds in ethnomusicological transcription and description. Particularly important in this book has been the exploration of the historical formation of the audile techniques used for disciplining the music–language relation in the global/colonial history of knowledge.

One of the tenets of Western music history is that the emergence of the work concept in Europe in the eighteenth century provoked an emancipation of music from language. But if we take into account the colonial archive then we can posit rather that the work concept emerged through the global circulation of discourses about music and language and that it was formed during the same historical period as the concept of orality. One of the theoretical proposals of this book is that the musical work and orality form a mutually constitutive pair that parceled out types of acousticities and political-expressive roles ascribed to different peoples in the late colonial period, easily seen in the differentiation between art and folk music. While the idea of the musical work was associated with the concept of transcendence, the idea of orality was associated with the notion of the people (or the subaltern classes) as central to enlightened nationalist politics. The notions of orality and of the musical work thus exist as a mutually constituted and constitutive pair by virtue of their simultaneous entanglement and differentiation in their role in the production of an idea of the person as an autonomous subject. But while in mainstream Euro-America the history of disciplinization of musical fields divided philology/folklore/linguistics from musical disciplines, this did not happen in Latin America and the Caribbean (or indeed in other parts of the world) where the dispersed disciplinization of the musical as a topic of study gave rise to different combinations of ethnology, the literary-linguistic-philological and the musical, several of which have been addressed in these chapters. As is the case in many cultures, the interrelated history of Western musics and language/orality share a fuzzy boundary zone that makes it hard epistemic labor to produce and maintain them as completely separate realms even when they can be identified as distinct phenomena (Feld et al. [2004] 2006; Faudree 2012).[1] But how that fuzzy boundary zone is conceived is not only a product of different cultural understandings of the relation between music and language. It is also a product of the way modern histories of knowledge have been globally constituted through colonial and imperial politics taking different shape in different parts of the world, in part as a result of modernist tendencies to differentially parcel knowledges out to different peoples.

In Latin America and the Caribbean the problems posed by the disciplinization of vocality in music and language have been historically tied to the tensions between metropolitan and Americanist belongings as well as to the rise of new discourses and genres as a response to the conflicts posed by colonial dominance and by the politics of modernization (Rama [1984] 1996; Ramos [1989] 2003; Martín-Barbero [1987] 2001; Hanks 2010). The categories of the lettered and the "popular" (as in *música popular* or in *cultura popular*) thus emerge as spheres of thought associated not solely with the rise of mass media (as in the notion of popular culture, in English) but they also reference the density of the political processes that determined their emergence in the entanglement of practices of vocalization, literary and philological inscription with the transformation of the sensorium. The "oral" then while simultaneously naming a disciplining practice of governmentality also names its own limit (Rama [1984] 1996); Ramos [1989] 2003). The methods for disciplining and governmentalizing a practice do not necessarily encompass all the manifestations and implications of such a practice. It is not just a question of opposition or resistance to them either (although, of course, they can take that form). Rather, the acoustic object *simultaneously* names the potential of its governmentalization and the difficulty of its proper containment, its tendency to come out of bounds. Its use for a particular politics and epistemology does not define its ground of existence.

But the voice was not only inscribed in writing throughout the nineteenth century. As we have seen in this book, the voice is also particularly amenable as an audile technique for bodily inscription through practices such as ritual possession as in Afrodescendant religions, trans-species musical pedagogies as in the indigenous practices of learning music from nonhuman entities or singing back to them, and in the use of vocal acoustics in the regimentation of the body as part of the vocal pedagogy of the nation-state. As such, the voice is a central acoustic dimension of the nexus and/or disjunctures between cosmology and political theology of the nation-state.

In this book we have explored the central role of practices of listening to the constitution of a political theology of the state that subsumed vocality to the politics of religion as a sphere of governmentality. We have also explored how different vocalities (such as vocalizations considered as irrational animal howls) found in the colonial archive can be used to understand the modes of manifestation of a cosmological order different from that proposed by the political theology of the nation-state. As such, listening appears located between the juridical, affective and everyday practices of the disjuncture between the theological and the cosmological, between the rise of a secular order and everyday secularisms, and between different understandings of the entities of "nature."

The differentiation and grounds of relation between nature and culture or, to name it differently, the relation between ontology and epistemology, between the ground of being and the figuration of knowing, was ultimately what was at stake in the disciplining of music and language. Today, the "return" of the importance of what had been historically understood during modernity as the separate sphere of "nature" to the constitution of the human itself has been formulated in many different ways by different areas of study. It has also been acknowledged in ethnomusicology, musicology, and composition as the emergence of acoustic ecology and, more recently, as ecomusicology, biomusic, and zoomusicology. Rather than describing or critiquing these emerging fields, a project that is beyond this book, I am more concerned with unsettling the very ground upon which these new, sub-fields of musical study are based.

As we have seen, what is crucial about the oral/aural relation to understanding the emergence of sound studies or, more broadly, as part of the genealogy of the acoustic sciences is that due to its location at the juncture of the nature-culture conundrum it confuses the history of the epistemes, of a radical differentiation between epistemology and ontology. As such, the epistemic transformation named by the emergence of sound studies is not an isolated event but part of the broader change of the relation between the human and nonhuman sciences, between ontology and epistemology, due to the contemporary irruption of "nature" as an unsilenceable political category in the affairs of the social sciences and the humanities. So what is happening when we talk about an "aural turn" is not necessarily that suddenly hearing has emerged on the scene today. As historians of sound increasingly show us, the archive tells us on the one hand, that many practices and disciplines central to modernity have to do with hearing and, on the other, that the history of globalization needs to take into account histories and understandings of listening that come from radically different regions, that point to different ontologies and politics of life and cannot be subsumed under the epistemic formations of Western modernity. Rather, what is crucial is that the changing relation between nature and culture regarding questions about hearing implies a reorganization of our own questions and disciplinary divisions around sound. Thus, the rise of "sound studies" as part of an ensemble of contemporary epistemic–ontological transformations in the relation between the sciences and the humanities is part of an economy of knowledge caused by the transformation of the historical relation between nature and culture in the recent history of the West. As stated by Isabelle Stengers, "in these new times we are dealing not only with a nature 'to protect' against the damages caused by humans but also a nature capable of disorganizing our knowledges and lives for good" (2009, 14). One of the crucial elements of unsettling the nature-culture relation in Western disciplines is its political urgency, manifesting for some as a "terrifying communication of the geopolitical and the geophysical" that "crumbles the foundational distinction of the social sciences, that between the cosmological order and the anthropological one, separated, for ever, that is, at least since the seventeenth century (let us remember the air pump and the Leviathan), by a double discontinuity, of scale and essence: evolution of species and history of capitalism, thermodynamics and stock exchange, nuclear physics and parliamentary politics, climatology and sociology—in two words, nature and culture" (Viveiros de Castro 2011a, 3).

The danger with the disciplining of each new field lies in curtailing its political significance as it enters the machine of the knowledge economy, which today is increasingly determined by new forms of juridization of life, of production, and of consumption. As we become more aware of the neurobiological processes involved in sound's capacity to affect bodies and senses, so do the persons who determine the practices of governmentality to which such knowledge is given use, as has been demonstrated by the use of sound as a vanguard military technology in the past few decades. The historiography of the relation between the ear and the voice, and more broadly of sound studies, then, is politically important because in highlighting that historical edge of sound, it simultaneously names the enactment and destabilization of the separation between the cosmological and anthropological orders, and the use of such potential for the implementation of different spheres of power. As such, the historiography of sound studies has the potential of giving political density to the current transformation of the relation between nature and culture, by naming the fragile limit between what sound simultaneously reveals and covers up, between cosmology and political theology as cosmology. The colonial history of vocality especially enables us to see the dangers of the juridical and lawful subsummation of cosmology into political theology as part of the eugenesis of the voice and the body. Thus the political question posed by the historiography of the voice is less about the acoustic intrusion of nature into culture and more about the consequences of such an intrusion (Stengers 2009). If naming is more about operating than about defining (50)—more a question of the types of action that naming sound studies enables than about determining the meaning of the field—that fuzzy edge of the limit between nature and culture that the historiography of the voice/listening nexus names becomes important for understanding the political edge of the current epistemic/ontological transformations taking place.

NOTES

All translations from the Spanish are by the author.

INTRODUCTION: *The Ear and the Voice in the Lettered City's Geophysical History*

1 See Jaramillo Uribe 1989; Silva 2005; Miñana Blasco 2000; Melo 2005.

2 See Hirschkind 2006; Seremetakis 1996; Rath 2003; La Tronkal 2010, among others.

3 Following historian Frédéric Martínez, I use New Granada only to distinguish it from the independent country, and I use the term *Colombia* as the name of the country even though this might not be historically accurate for the nineteenth century: "The first republic of Colombia (known as Gran Colombia) regroups the territories known at the end of the colonial era as Virreinato de la Nueva Granada, Capitanía General de Venezuela, and Audiencia de Quito. In 1830 the Republic of Colombia is fragmented into three republics: Venezuela, Ecuador, and the Republic of New Granada, which comprises the present-day territories of Colombia and Panama. This name is kept until 1858, when it is replaced by that of Confederación Granadina, which is itself replaced by that of Estados Unidos de Colombia. The 1886 Constitution reinstates the name of República de Colombia, which is used until the present" (Martínez 2001, 31). See also Tirado Mejía ([2001] 2007, 8).

4 Such a distinction, which assumes a passive nature upon which a politics is inscribed generating as it were a humanly politicized nature, has also been questioned, with a different vocabulary and approach, by the history of science (Stengers 2009; Latour 1993), by recent developments in the anthropology that question the category of nature (Escobar 2008; Viveiros de Castro 2010), and by the "speculative turn" in philosophy (Bryant, Srnicek, and Harman 2011).

5 The historiography of the colonial era and the early postcolonial period have been articulated by differently positioned scholars in deconstructionist projects in different moments, giving rise to what has been controversially labeled as Latin American cultural studies as well as to projects located outside of this purview in anthropology and history. This is part of an ongoing history of debates on the nature of modernity and of the colonial in the region. One of the debates of the colonial in Latin America and the Caribbean that has gained more visibility recently in certain countries in Latin America and in the United States has taken place under the aegis of what has come to be called the "modernity/coloniality group" or the "Latin American coloniality group." For a summary of their trajectories and positions, see Castro-Gómez and Grosfoguel 2007; Moraña, Dussel, and Jáuregui 2008. For a general introduction to Latin American cultural studies in English, see Del Sarto and Trigo 2004.

6   The centrality of the Americas to the imbrication of the human as a biological and cultural agent of history has been explored by Crosby (1986) who, in his environmental history, has shown the role of humans in the transformation of species and the environment in the region.

7   For a relationship between Alexander von Humboldt, Latin American and Caribbean naturalists, and the Hispanic American colonial archive, see Cañizares-Esguerra 2001; Nieto Olarte 2007; Serje de la Ossa 2005; Bunzl and Penny 2010.

8   For a detailed history of Western ocularcentrism, see Jay 1993; Crary 1992.

9   Hannah Arendt, after Whitehead, called such a philosophical positioning the Archimedan point of view. She did not link it to the colonial but to a process of despatialization of philosophy through the interrelationship between political theology, new visual technologies, and science that emerged in the Renaissance and that prevented it from recognizing the pluralism of the subject (Arendt [1958] 1998).

10 See Szendy (2007, 2008, 2009) for a critique of philosophy as ocularcentric; Schmidt (2000) and Corbin (1998) for a complex history of the sensorial and its relation to the secular-sacred; Hirschkind (2006), Seremetakis (1996), Feld (1996), Howes (2004), among others, for a critique of the relation between modernity and history of the senses.

11 I will not signal particular works here since this involves a deconstruction of the history of the emergence of Latin American and Caribbean aurality in the media and in folkloristics throughout the twentieth century and thus goes beyond this project.

12 He does explore the relationship between specific scholars of German comparative linguistics and the rise of Nazism (see Esposito [1998] 2009). But the concept of orality is central to the constitution of the idea of community and it is this particular aspect that he does not explore.

13 See Jonathan Crary (1992) for a similar turn to an existential "observer" and notion of observation in the nineteenth century.

14 That is why in ethnographies of the concept of the person in Amazonia, for example (Viveiros de Castro 2002; Stolze Lima 2005), there are abundant references to hearing as the entry point that provided an alternate understanding of the idea of the person, even if hearing itself is not theorized.

15 As Jonathan Sterne has shown, the notion of transduction is itself culturally constituted (2003). We can think of transduction as a process in terms of an understanding of the ear through ideas of engineering, of physiology, or of biochemistry. But no matter how we conceive of it at particular historical moments and locations, the important element to emphasize is the very transformation of energy that the relation between sound object and listening subject implies as an idea with which to think analytically.

16 To see how the very idea of culture is constituted through such equivocation, see Wagner ([1975] 1981).

Chapter 1: On Howls and Pitches

1   Humboldt began his travels in Latin America and the Caribbean in 1799. When he arrived in Cartagena on May 28, 1801, he had already traveled extensively in Venezuela and Cuba, where he set sail for mainland South America in order to meet the expedition of Nicolas Baudin in Lima.

2 All translations from Spanish are by the author.

3 For a basic understanding of how this process of perception is bodily constituted, see chapters 12–16 of Goldstein, *Blackwell Handbook of Perception*, 2001. See also Yost 2001.

4 Goodman links this to a mode of knowing he calls "politics of frequency," an interdisciplinary endeavor constructed as a "nonrepresentational ontology of vibrational force" (2010, xv). My method is different. Attention to the materials of the postcolonial archive require simultaneous attention to diverse understandings of transduction (Sterne 2003) and acoustemologies (Feld 1996, 2012).

5 Humboldt describes the champán and the bogas' rhythmic activity quite vividly:

"The form in which these bogas work is very rhythmic. While half of them (three, for example) move towards the roof with the pole supported against their chest, the other three move in the opposite direction, with their arms above their heads (holding the poles horizontally above the heads of those that are working) and moving towards the other extreme of the champán. While half of them reach the extreme of the boat, and while the others reach the other extreme, in that moment they put the pole in the water, and the others agitate the pole in the air and the champán, in this eternal going back and forth, can never gain time to slip back down with the current. In this way the bogas alternate each other on top of the roof" (Humboldt 1801, 27).

6 See, for example, Pratt 1992; Nieto Olarte 2000; Cañizares-Esguerra 2001.

7 Many more testimonies of bogas singing can be found in the documentation written by Creoles and Europeans in this period. This is a selection since other testimonies tend to emphasize the same issues.

8 For a detailed discussion of Hanslick's work, see Payzant 2002.

9 He links consonance with an immanent acoustic property based on the degree of fusion of two primary sounds. Fusion is defined as "the approximation of a two note clang towards unison . . . the more consonant a two note clang, the more the primary tones fuse, the more they approximate toward the sensation of unison" (Révész 2001, 88). Thus, the "two tones are perceived as a fused entity, depending on integer ratios, beginning with the octave and going through the intervals in four further categories (fifths, fourths, major and minor thirds, and all other intervals) in descending degrees of fusion" (Rehding 2000, 353).

10 See Rehding 2000, 2003; Moreno 2004; Clark and Rehding 2001; Steege 2012; among others.

11 Rehding highlights that the theories of musical origin that prevailed in this period stemmed from British "positivism": Darwin's theory of music originating in bird songs for sexual selection; Spencer's theories in which music originated in speech, and Karl Bucher who proposed that music originated in the "repetitive motions of physical labor" (Rehding 2000, 350). Carl Stumpf explained these and other British theories of musical origin in his *Musikpsychologie in England* (1885). For a summary of the different theories that appear in this text, see Elizabeth Valentine, "Carl Stumpf and English Music Psychology" (2000); for theories on the relation between the origins of music and language in the French Enlightenment, see Thomas (1995).

12 For the relations between the bogas as laborers and the emergence of a working class in the second half of the nineteenth century, see Solano (1998).

13 In the sixteenth century, a legal battle ensued between the Crown, missionary priests, and the *encomenderos* of Mompox regarding the treatment of the Indians in the boga. This was marked by a succession of royal decrees and letters to the Crown in the mid-sixteenth century that tried to regulate the work of the Indians (August 11, 1552, August 14, 1556, June 24, 1561). But such decrees had no weight in the midst of the open rebellions of the encomenderos against the Crown's regulations and of the Indians against the encomenderos. The Indians then "were dying like flies," as described in a letter by Martín Camacho, functionary of the Crown, to the King, written in Cartagena on July 10, 1596. He states: "that he has seen that there is no year in which the boga does not consume at least 500 indians" (Camacho in Noguera Mendoza 1980, 68). By the time the legal battle was resolved in the 1560s, the Indians of the boga in the Magdalena River had been killed through forced labor or disease, had "chosen" suicide over slavery, had retreated to the extensive frontier and unpoliced lands of the Colombian Caribbean, or mixed with other populations.

14 Cartagena was the largest slave port in South America between 1580 and 1640. See Maya Restrepo (2005).

15 According to this same census, "civilized indians" made up 22.1 percent of the rural population in the region (meaning there were more in the hinterland), "there were no more than 7,708 slaves outside the six principal cities in the 1770s amounting to 5.7% of the rural population," and "whites residing in the small villages and towns amounted to 9,898 people or 7.3% of the non-urban population, that is, slightly more than slaves" (Helg 2004, 46–48). At the urban level, of a total population of 35,051 in Cartagena, the majority were free persons of all colors (56.8 percent), 27 percent were white, and 15.7 percent were slaves. The second-largest important city, Mompox, according to this same census, had the largest proportion of free people of all colors (74.3 percent), with whites making up 12.9 percent of its inhabitants and slaves 11.7 percent, for a total population of 7,197 inhabitants. In terms of this census, peoples were administratively classified in different ways. Says Helg, "Only the census of Riohacha Province distinguished between mulattoes, zambos, negros and mestizos among the free people of color. . . . The censuses of Cartagena and Santa Marta provinces did not classify the free population of color in distinct racial categories" (2004, 43).

16 For different interpretations and a detailed, comparative discussion of this census and other numbers regarding the Atlantic slave trade in New Granada see Colmenares [1973] 1991; Jaramillo Uribe 1989; Cunin 2003; Maya Restrepo 2005.

17 This included factionalism between the elites of the different cities and divisions between those that supported Bolívar or the 1810 Constitution. Colombia declared its independence from Spain in 1810. The years between 1810 and 1816 were marked by intense battles and factionalism and are known as the Patria Boba (the dumb country). The Spanish regained control of the territory in 1816 installing a period known as the Regime of Terror. In 1819 Simón Bolívar and his armies marched into Bogotá regaining control of the territory and the first constitution was written in 1821 amid great division among elites from different regions.

18   See, for example, the detailed cases of the Palenque de Matuderé described by Jane Landers (2002) where captured indigenous and Spanish women were assigned as wives of escaped ex-slaves (cimarrones).

19   The Inquisition was established in Cartagena on February 5, 1610, and lasted until 1819, when it ended due to Colombia's independence from Spain. The other two seats of the Inquisition in the Americas were Mexico City and Lima.

20   This is confirmed by Adriana Maya Restrepo who states that after 1720, the juridical processes that judged this type of crime practically disappeared in New Granada (2005, 504).

21   See Restrepo 2005 for a critique on Huellas de Africanía; see Arocha 2004 and Maya Restrepo 2005 for a support of this position.

22   See Basso 1985; Guss 1989; Hill 1993, 2009; Stolze Lima 2005; Montardo 2009; Reichel-Dolmatoff 1991; Seeger 1987; Viveiros de Castro 2004.

23   For a detailed genealogy of the concept and its problems see Bird-David 1999, including responses to the article.

24   Either as a celebration of an acoustic ecological sensibility (Velasco) or as decolonial critique of his sensorial acuity (Kueva), Humboldt's oeuvre appears as one that anticipates the idea of acoustic ecology. To explore Daniel Velasco's pieces, see his own description of his work in Velasco 2000. Fabiano Kueva's sound intervention is found at http://www.pangaea-mq.com/español/artistas/fabiano-kueva/.

25   For an elaboration of how this happened, see Nieto Olarte 2000; Castro-Gómez 2004; Pérez 2004; Cañizares-Esguerra 2006.

26   Humboldt's persona and work have been used to uphold multiple political projects, generating interpretations of his work in the service of completely different ideologies. See Rupke 2008 for this "metabiographical" history. See Cañizares-Esguerra 2001, 2006; Nieto Olarte 2007, 2010; Serje de la Ossa 2005, for discussion on his colonial stance.

Chapter 2: On Popular Song

1   Cauca is Isaacs's native state and where the novel *María* primarily takes place.

2   The struggle between federalist and centralist models of governance and between regional, national, and cosmopolitan allegiances of different elites gave way to many forms of political conflict. According to Alvaro Tirado Mejía, "during the nineteenth century, besides the many local rebellions, there were eight major civil wars: the one between 1839 and 1841, known as the War of the Convents or the Supremes; that of 1851; that of 1854; that of 1859 to 1862; that of 1876–77; that of 1884–85; that of 1885; and that of 1899–1902, known as the War of One Thousand Days" (Tirado Mejía [2001] 2007, 8).

3   Lázaro Mejía Arango divides the period of dominance of radical liberalism into two stages: from 1864 to 1874, in which radical liberals could execute their political reforms with success, and between 1874 and 1886, where they became divided and eventually were defeated by the consolidation of the Conservatives in power in 1886 allied with liberals who had abandoned the radical stance. See Mejía Arango 2007, for a detailed political history of the radical liberal presidencies.

4   The Conservative Party was officially created in Colombia in 1848 and was closely allied to a project of a national Catholic state. Along with the Liberal Party, it has been one of the main two parties of the country.

5   As Martín-Barbero brilliantly summarizes it, this was grounded on the different "translations" that accrued in the term *popular* in the eighteenth century: popular as articulated by the French revolution, the expression of a people idealized in their cultural potential yet rejected as a participatory political figure; popular as the geopolitical intention of identifying nationalistic claims via folk song and language as rooted in particular ideals of community in German Romanticism; popular as signifying a temporal orientation to the past through evolutionary theory and ideas about heritage, developed through the impulse of antiquarianism of Scottish and British intellectuals; popular as the expression of the emotional and spontaneous (and potentially as something that is not true) contra Enlightenment's rationalism; and popular as signifying the revolutionary potential of the people through its relation to the proletariat as portrayed in Spanish anarchism (Martín-Barbero [1987] 2001).

6   I owe this insight, above all, to discussions with my students who work on song craft in different parts of the world. Thanks especially to Farazaneh Hemmassi (on Iranian popular music), Lauren Ninoshvili (on vocables in Georgian music), and Adam Kielman (on Confucianism, state ideology, and folk song in China). For anthropological studies of how song is understood and sensed as producing this relation in genealogies that are not those of eighteenth-century Romanticism, see Samuels 2004; Feld 1996; Meintjes 2005; Seeger 1987; Erlmann 1999; Fox 2004, among others.

7   The Cancionero was in the library of Isabel la Católica, Queen of Spain, from where it moved to El Escorial, to be taken from there and bought by the Bibliothèque Nationale in France, where it was found by diplomat and medievalist Pedro José Pidal in the nineteenth century. Pidal published it in Madrid in 1851 with a highly influential prologue. The extant copy is not the original one. See Weiss 1990.

8   It is impossible to know exactly who Vergara y Vergara is drawing from when he refers to Spanish medieval poetry due to the loose practices of (non)citation that characterize his work and this period in general. The multiple parallelisms between José Pidal's nineteenth-century introduction to the *Cancionero de Baena* and Vergara y Vergara's text, and a book he mentions in the introduction, the *Resumen Histórico de la Literatura Española* by Don Antonio Gil de Zárate, are probably two of the most influential books from which he draws his narrative. Though the *Cancionero* was a highly influential work, it was not the only Renaissance songbook published in the mid-nineteenth century. These included the works of the Marqués de Santillana, compiled and published in 1852 in Madrid, as well as the works of Renaissance poet Enrique de Villena along with several compilations that in the nineteenth century sought to reedit the fifteenth-century cancioneros. One can also find compilations of popular songs such as the songbook Vergara y Vergara mentions in his final chapter, that of Emilio Lafuente y Alcántara, published in Spain in 1865, among others. The fact that Vergara y Vergara probably had access to these books is evidenced by their existence, in many cases more than one copy, in the Biblioteca Nacional and other archives in Bogotá, meaning they at least

circulated in private collections of erudite persons in Bogotá and probably beyond that. But regardless of which texts Vergara y Vergara references specifically, what is crucial, in any case, is that he was very much aware of the foundational and epic tones of the nineteenth-century redeployment of such Spanish literary origins.

9   The Muiscas are the indigenous group that inhabited the plains of Bogotá, and Chibcha is the language family to which they belong. In the chronicles these two terms are used interchangeably.

10 In the nineteenth-century prologue to the *Cancionero de Baena* and in the nineteenth-century literary histories of the Spanish language, a clear distinction is made between songs that narrate history but are not sophisticated in style and poetics and songs that are about love but are highly sophisticated in their poetics. The former is the art of the *juglares*, low class poet-singers who worked for pay, the latter, the art of the troubadours, courtesan poets (Pidal 1851).

11 Colombia officially recognizes today the existence of sixty-five indigenous tongues belonging to "more than eleven linguistic families." See http://www.lenguasde colombia.gov.co/content/lenguas-de-colombia.

12 As Ana María Gómez-Bravo points out, the history of the relation between songs to be sung (*cantares*) and songs to be recited or told (*decires*) in medieval Spain does not necessarily rest solely on their lack or presence of musical accompaniment. For her the mid to late fourteenth century marked a transformation in the understanding of song genres due to the rise of Renaissance polyphonic chant, which required specialized professional singers (rather than simply a popular singer accompanying him- or herself), and changes in the practices of reading, from reading out loud to reading silently. The difficulty of the distinction then rests on a complex history of transformation of such a relation and on the rise of a rhetorical theory that speaks of different modes of enunciation (Gómez-Bravo 1999). However, the difficulty of clearly distinguishing between "cantar" and "decir" is, of course, not only a phenomenon of Spanish medieval poetics. In medieval chant "cantare officium admits the double translation of 'to sing' or 'to say' the Office. As a matter of fact, this ambiguity is consequent upon the ancient and medieval practice by which plain reading, the "lectio plana," was always done aloud. Silent reading was unknown and even private recitation of psalms was always done aloud. The medieval reader understood by listening to himself. Hence, in his mind the differences between *legere* and *cantare*, in so far as the voice was concerned, were only differences of degree, i.e. of vocal intensity and intonation" (Dijk 1952, 8). Ultimately, then, this ambiguous vocality between speech and song emerges historically there where the question of the relation between voice and inscription requires attending to the disjuncture between the voice and the disciplinary technologies that seek to contain it.

13 In Colombian twentieth-century folkloristics, the bambuco became the national genre par excellence, contra the ascendancy of the cumbia as the most popular genre that arose via the music industry and that came to be identified with the Colombian nation both in the Caribbean and abroad. This disjuncture between modes of presenting the nation in the early twentieth-century history of musical nationalisms in Latin America is unique and ultimately speaks of the histori-

cal division between the Caribbean and the Andean regions in terms of cultural and political representation. The disjuncture is most clearly manifested in the division between a grammarian nation, represented by Bogotá, and a Caribbean nationhood that makes its full appearance in the transnationalization of Colombian "música tropical" and in García Marquez's *One Hundred Years of Solitude*. It ultimately speaks of the ways that Colombia is primarily understood, until today, as a country marked by a divisive geography.

14 The topic of the African ascendancy of the bambuco, understood as something that was reflected in its polyrhythmic and polymetric character, was intensely debated by folklorists and composers in the first half of the twentieth century in Colombia. See Santamaría-Delgado 2014.

15 As explored by Viveiros de Castro, in the conquest of the Americas, "the ethnocentrism of the Europeans consisted in doubting that the bodies of the others contained a soul that was formally similar to that which inhabited their own bodies; the ethnocentrism of the Indians, on the contrary, consisted in doubting that other souls or spirits could be outfitted by a materially similar body to that of indigenous bodies. In terms of Roy Wagner's semiotics . . . the body would belong to the dimension of the innate or spontaneous in European ontology ('nature'), a dimension that is the counter-invented result of a 'convencionalizing' symbolic operation, while the soul would be the constructed 'differentiating' dimension." (Viveiros de Castro 2010, 29).

16 Emilio Lafuente y Alcántara (1825–1868) was a historian, archivist, and collector of medieval, Arabic, and popular traditions in Andalucía, Spain. The second edition of his songbook was published in 1865. In the prologue to this book, Lafuente y Alcántara establishes a close correlation between song, passions, "characters and customs" (vii), and song types. The point is to show the circulation of ideas and books between Spain and New Granada.

17 For the "ordering of nature" by the illustrated men of letters in New Granada, see Nieto Olarte 2000, 2006, 2007; Castro-Gómez 2004; Serje de la Ossa 2005, among others.

18 For different interpretations of this controversy, see Cañizares-Esguerra 2006; Nieto Olarte 2007; Serje de la Ossa 2005.

19 This is another long historical debate in the Americas regarding the nature of the interpretation of Spanish chronicles in Europe from the sixteenth to the late eighteenth century. See Cañizares Esguerra 2001. The debate was recast anew by patriotic Creoles such as Caldas vying for independence from Spain. Caldas became one of the New Granadian martyrs of independence when he was killed in 1816 by Pablo Morillo, head of the forces of the Spanish Reconquest sent to New Granada between 1815 and 1816.

20 Vergara y Vergara himself edited a volume entitled *Museo de Cuadros de Costumbres* (1866), a compilation of Colombian writers working on the genre. It was also cultivated by some of the scholars of the Chorographic commission (1849–58), a geographical project meant to map the country. Several of the members of the commission became part of El Mosaico and made the diffusion of their work one of its central projects (Von der Walde 2007).

21 The change of discourse from disgust to a racialized celebration of Afrodescendants in song "as a gift" is actually quite notable when compared to earlier
chronicles.

22 Elisabeth Le Guin (2006) calls "carnal musicology" the recuperation of the relation between performer and composer through a historical reinterpretation of musical form, embodiment, theory of affect, and the senses in the work of Boccherini.
Palmié (2002) calls "forms of carnal knowledge" the logic of production of othered
bodies in the history that extends from Caribbean slavery to contemporary sexual
tourism in the region. Acknowledging the place of the master-slave narrative in the
history of the body and its relation to music would be, perhaps, the crucial feminist
deployment to undertake here, not simply adopting an ontology of musical vibrations as pleasure.

23 Carlos Jáuregui enumerates his works between 1878 and 1884 "three translations
of a manual of military tactics (Nociones de táctica de infantería, de caballería
y de artillería, 1878), an amorous memorial written in prose, original poems and
translations (Lectura para Ti, 1878), and amorous and moralizing drama on social
customs (Secundino el Zapatero, 1880), a fragmentary poem-drama about a mis-
understood poet who is in love (Lucha de la vida, 1882) and three translations and
adaptations to Spanish of Italian courses (1883), French courses (1884) and English
(1884)" (Jáuregui 2007, 47). *Cantos* is then an early piece.

24 "The initial r has the smooth sounds of the intermediate r in those words (*voces*)
in which it replaces the d. The c sound is strong in words such as these: libectá,
ficmeza. That of the articulation j, when it replaces s, is extremely brief and
somewhat imperceptible. E stands as ej (es) and many times re (de) especially in
composed words (lengua-e-vaca), and also when it is required by the elegance of
the phrase or the structure of the word. Er (pronounced éer) is equivalent to der
(del) and is different from er (el) as much as opposed quantities stand apart from
each other. To establish this difference on the written page, I mark this sign over
that voice thus: ér. That ér is as valuable as dér, cannot be doubted" (Obeso [1877]
2005, 43).

25 The similarities between Obeso's most famous poem "Canción del Boga Ausente"
and this *bunde*, have been repeatedly noted: *se no se junde la luna; /Remá, remá./
Qué hará mi negra tan sola /Llorá, llorá /Me coge tu noche escura /San Juan San
Juan./Escura como mi negra,/ni má, ni má/La lú de s'ojo mío/ Der má, der má./Lo
relámpago parecen/Bogá, bogá*. And a fragment of Obeso's poem: *Que trite que etá
la noche,/La noche que trite etá/No hai en er Cielo una etrella . . . /remá, remá/ Ra
negra re mi arma mía/ Mientrá yo brego en la má/Bañaro en suró por ella, /Qué hará,
qué hará?*

26 He wants to love or woo a white woman but cannot do it because of his color
(Oh! branca, branca hecmosa/ Pocqué me trata asina?/No sabe que la ejgracia/
Re compasión e rigna? [Oh beautiful white woman/ why do you treat me thus?/
do you not know that disgrace/ deserves compassion?]); he has been a soldier
in the armies of the nation against the Conservative party but refuses to become
a Caribbean soldier against Andean Bogotá yet again (quieren la guerra con los
cachacos?/ Yo no me muevo/re aquí e mi rancho. [Do you want war with the ones

from Bogotá?/ I will not move from here/from my hut]); He counters the association of blackness with ignorance and lack of respect: (Cuando soi un probe negro, /sin ma cencia que mi oficio/ No inoro quien se merece/Acgun repeto y cariño. [When I am a poor black man /with no science but my trade/I do not ignore the person/that deserves some respect and kindness]).

27 Isaacs's detailed literary career and his broader politics of listening are the topic of the third chapter.

28 See Santamaría-Delgado (2007, 2014) for a summary of some of these debates and Miñana Blasco (1997) for the cartographic history of homogenization of the genre.

29 Isaacs had to abandon his job as road inspector due to malaria, the disease from which he would eventually die, which he acquired during this period in the Dagua.

30 See Colmenares (1976, [1973] 1991) and Almario (2005) for a history of the Cauca region. See Escobar (2008) for an anthropological history that links the Cauca to the broader Pacific region.

31 For a summary of these perspectives up to 2004, see Avelar (2004).

32 See next chapter for Caro's critique of Isaacs's Jewishness.

Chapter 3: On the Ethnographic Ear

1 Uricoechea states that he will leave for a later time, after having edited the indigenous grammars of New Granada, "the two big questions that philology has dealt with, the origin and history of the human species as revealed in its tongues" (1871, xii). Nevertheless, the introductions to the three indigenous grammars he edited, his correspondence, and his work in general are steeped in questions about such origins and history.

2 For the philosophical perception of an opposition between the linguistic turn and a speculative turn to the real see Bryant, Srnicek, and Harman (2010, 1–18).

3 These life history details are mostly taken from the excellent biographical sketch of Uricoechea done by Clara Isabel Botero (2002). Her work brings together important biographical elements done by different scholars and is the most systematic and complete to date.

4 He wrote to Rufino José Cuervo in no uncertain terms about his experience in Colombia: "He hecho muchas cosas en mi vida, pero la mayor bestialidad de todas fue irme a meter de cabeza en bogotá en tiempos de libertad golgótica en que nada se hacía por la instrucción. En fin, no dejé el pellejo y debo considerarme feliz" (letter to Cuervo, September 5, 1876, in Romero 1976, 176–77). And again in the prologue to his Mapoteca (1860): "Sepultado en el centro de los Andes, sin eco mi voz, sin estímulo á mi alrededor, había abandonado la idea de hacer publicación alguna" (Uricoechea 1860, vii).

5 For detailed accounts of missionary grammars in the Americas, see Hanks 2010; Zwartjes and Hovdhaugen 2004; Zwartjes and Altman 2005; Zwartjes, James, and Ridruejo 2007.

6 According to Triana y Antorveza (1987), José Celestino Mutis, the Spaniard who led the New Granada Botanical Expedition, was put in charge of collecting grammars for this collection by express orders from the Spanish monarchy. In his introduction to the Páez vocabulary Uricoechea lists the materials sent to Catherine the

Great by Mutis: "Como interesa conocer los antiguos manuscritos de las lenguas americanas, aprovecho la ocasión para dar la lista de los que envió Mutis en aquel entonces para Rusia y cuyo paradero ignoro: 1. Gramática, Vocabulario i Confesionario de la lengua Mosca-Chicbcha. El orijinal está en poder del Sr. Mútis; se presume escrita por Dadey; 2. Vocabulatio Mosco, más antiguo que el anterior; 3. Arte i vocabulario de la lengua Achagua, compuesto por los que trabajaron les [sic] Pp. Alonso de Neira i Juan de Ribero; 4. Vocabulario de la lengua que usan los indios de estas misiones—del Colejio de Popayán—se supone ser la lengua de Ceona; 5. Diccionario de la lengua Andaquí de dichas misiones; 6. Idioma de la provincia de Páez sacado por Eujenio del Castillo con un pliego de voces del idioma de la Nación Murciélaga o Huaque; 7. Traducción de voces castellanas de la lista año 2 (mandada de San Petersburgo en lengua Motilona i breve dicccionario español Motilón; 8. Catecismo para información de los indios Coimas, Sabiles, Chaques i Anatomos; 9. Traducción de las voces de la lista no. 2 en lengua de los indios Guamos; 10. Traducción de la segunda lista en lenguas Otomoaca, Taparita y Yarura; 11. Breve compendio de nombres sustantivos I adjetivos o términos comunes I necesarios para estudiar la lengua Pariagota; 12. Frases I modos de hablar traducidos en lengua Guarana; 13. Vocabulario escrito para la lengua Aruaca (Uricoechea 1877, xx–xxi).

7   Wilhelm von Humboldt's knowledge of American languages, for example, came from his personal acquaintance with Hervás, whom he had met during his stay in Italy between 1802 and 1808, and his familiarity with Hervás's work, even though he denigrated the old priest's attitude toward languages as unscientific (Tovar 1986). According to Triana, P. S. Pallas and Jiankiewitsch de Mireo, who prepared the comparative vocabularies for Catherine II of Russia, did not know Hervás's collections (Triana 1987, 15). That is why they looked for American languages directly in the Hispanic-American colonies, by express order of the Spanish Crown, in response to the request of Catherine the Great. See Triana 1987. Breva-Claramonte (2001) also points out Hervás's clear contributions to issues of the typology of languages in the eighteenth century, particularly due to the fact that geographic proximity played a minor role in his comparisons because of the enormous amount of material he had available from around the world. He therefore relied more and more closely on linguistic analysis in his comparisons, an aspect that is usually considered crucial to the modernization of philology.

8   On March 17, 1876, Rufino José Cuervo wrote a letter from Bogotá to A. F. Pott in which he sent him, through Ezequiel Uricoechea, his book *Apuntaciones críticas sobre el lenguaje bogotano* (Critical notes on Bogotano language) (1867), with hopes that the book might be useful to him "especially because in it are found many of the corruptions and alterations that have occurred to one of the most beautiful modern tongues due to its isolation in a distinct region. Maybe some analogy, be it phonetic or lexical or syntactic can open for you vast horizons . . ." (Cuervo, in Schutz, ed., vol. I, 1976, 195). Cuervo deemed that his particular study of the specifics of a language in a remote location could contribute to the larger comparative enterprise, through its inclusion by Pott. Pott replied kindly to this letter and gift, thanking Cuervo, expressing his immense happiness for seeing that philology

was being cultivated in such far-off latitudes as Bogotá, and engaging him in a debate about the need to understand the "corruption of language" as the inevitable historical transformation of languages (Pott, letter to Cuervo, Halle, June 9, 1876, Schutz, ed. vol. 2, 1976, 245–50).

9 Uricoechea takes this latter idea directly from Jesuit priest Antonio Julián, a missionary among the Goajiro in New Granada, and from his "prophetic book *The Paradise in America* in which he tries to prove, Bible in hand, that Adam and his descendents set out from Colombia and whose original manuscript we possess" (*Uricoechea el Americano*, París, June 11, 1872, 203). He enthusiastically cites this article for Juan María Gutiérrez, stating that "the idea is not new, this much I can prove with Father Julián's manuscript, but the mode of presenting it merits study" (Uricoechea, letter to Gutiérrez, August 3, 1872 [1998], 118). The Andes as paradise had a long history in the region by the time of Uricoechea. See Cañizares-Esguerra 2001, 2006.

10 Or using Connor's language — mishearing is brought from nonsense into sense by the ear.

11 Much of the close analytical work with missionary grammars has been done with those of New Spain because of the abundance of such grammars in this region.

12 See Hanks 2010; Zwartjes and Hovdhaugen 2004; Zwartjes and Altman 2005; Zwartjes, James, and Ridruejo 2007 for such a discussion. My idea, which does not appear in these discussions, is that the model for such descriptions of indigenous languages is chapter 5 of Nebrija's Spanish grammar in which he describes the problems of Spanish language as a foreign tongue. Uricoechea, however, states that the model followed by all such grammars was "la vascuence de Larramendi que vino a alterar el canon universal moderno" (1878, xlv).

13 Whether such a physiological *and* explicatory supplement is understood as the Derridean aporia of the impossible to solve deferral of proper acoustic articulation *and* signification, as a Deleuzian or Whiteheadean open-ended becoming, an "incessant novelty, and 'perpetual perishing' [that] do not make reference and grounding impossible" (Shaviro 2009, 150), or as a Lacanian chain of signifiers through which desire emerges, depends in good measure on the way such a disjuncture is woven into the experience and understanding of the politics and poetics of uncovering American singularity. Moreover, Patrice Maniglier's detailed reinterpretation of Saussure's work complicates how the problem of sound is addressed in Saussure. For him, this is one of the reasons for rethinking the relation between the linguistic and anthropological dimensions of the structuralist project and its relation to the French philosophies of difference. See Maniglier 2006, 2011.

14 I would like to thank Steven Shaviro for noting the connection to the Sapir-Whorf hypothesis.

15 I am not implying a genealogical connection between Uricoechea and these developments, simply a narratological one in the sense of resources available for theoretical and aesthetic deployment.

16 The battle of 1876, which the radical liberal won, is considered decisive in their eventual defeat. Radical liberals had dominated throughout the century. But this battle became an event that exposed the crisis of their government, and the begin-

ning of the realignment of forces between moderate liberals and conservatives that ultimately led to the rise to power of the conservatives. See Mejía Arango 2007.

17 One of several examples: "Hase convertido en un empleo lucrativo y honorífico la representación de indígenas: las cábalas y las prestidigitaciones en la farsa del su-fragio popular, son también de provecho y usanza en las regiones salvajes del país. Afirmo, sin riesgo alguno de que se me contradiga, que en los territorios a que me refiero jamás hubo un centenar de indígenas salvajes que supiera de qué se trataba en las tales elecciones de Comisario; y sé que los aborígenes bárbaros no conocen ni de nombre a las personas que vienen a representar los intereses de esas tribus en el congreso nacional" (Isaacs [1884] 1959, 167–68).

18 During the Conservative hegemony at least four of the presidents were studious philologists and grammarians: Miguel Antonio Caro (president 1892–98), José Manuel Marroquín (president 1900–1904), Marco Fidel Suárez (1918–21) and Miguel Abadía Méndez (1926–30). See Rodríguez-García 2010.

19 The Constitution was officially signed on August 5, 1886. This was preceded by a discussion among persons designed to represent the different states (delegatarios) throughout 1885 and 1886. For a detailed account of this discussion, see Valderrama Andrade 1997.

20 The idea of sin penetrated political discourse and party affiliations to such an extent that General Uribe Uribe, one of the few liberals and opponents of Caro in congress during the 1890s, felt compelled to write in 1912 a booklet entitled *De cómo el liberalismo politico colombiano no es pecado* (On how Colombian political liberalism is not a sin) ([1912] 1994).

21 Humboldt's study in turn is based on Father Duquesne's *Disertación sobre el Calendario de los Muiscas,* an earlier study on Muisca pictographs. This was given to Humboldt by José Celestino Mutis during his stay in Bogotá (Pérez Arbeláez 1959, 176). Humboldt used it extensively on his own writings on the Muisca, basically rewriting and copying it and simply adding a comparison to Asian mythologies and archaeological figures, something admitted by Humboldt himself (177).

22 Two examples suffice: "Si notoriamente falta en la población la unidad de raza, y en el territorio la unidad de topografía y clima, al contrario, por lo tocante a la re-ligión, como al idioma, la unidad social es completa. De aquí la necesidad y la jus-ticia de reconocer a la religión única del pueblo colombiano" (José María Samper, 349, Derecho Público Interno de Colombia, cited by Rodolfo Arango 2002, 139). And another example, "De suerte que mirar por la lengua vale para nosotros tanto como cuidar los recuerdos de nuestros mayores, las tradiciones de nuestro pueblo y las glorias de nuestros héroes y cuando varios pueblos gozan del beneficio de un idioma común, propender a su uniformidad es avigorar sus simpatías y relaciones, hacerlos uno solo. Por eso, después de quienes trabajan por conservar la unidad de creencias religiosas, nadie hace tanto por el hermanamiento de las naciones hispano-americanas, como los fomentadores de aquellos estudios que tienden a conservar la pureza de su idioma, destruyendo las barreras que las diferencias dialécticas oponen al comercio de las ideas" (Cuervo 1867, 6).

23 Also, in the early twentieth century, the development of linguistics of indigenous languages would take place in Colombia, largely under such missionary tutelage.

See especially the work of Father Marcelino Castellví, a Capuchin priest, founder of CILEAC (Centro de Investigaciones Lingüísticas y Etnológicas de la Amazonia Colombiana).

Chapter 4: On Vocal Immunity

1   There are several parallelisms between Aristotle's and Caro's notion of the voice. While I call attention to the parallelism, elaborating its full implications is beyond this text. Aristotelian ideas about the animal in the voice as well as about the differing power of consonants and vowels were central aspects of Caro's work discussed in this chapter.

2   The first part of the treatise on orthology consists of the explanation of the pronunciation of the Spanish language, in which discrete articles such as vowels, consonants, or syllables are explained in terms of their sonic, bodily output: "b is pronounced with the two lips and v with the top teeth and inferior lip" ([1869] 1874, 3). The second part consists of the problems of a proper accent, defined as "the effort made over one of the vowels of each word, giving it a stronger tone or elongating the time it is pronounced" (5). This is followed by a series of rules of proper accentuation. The third part of the treatise consists of a regimentation of the problem of accentuation in words that have consecutive vowels, diphthongs, and triphthongs, followed by a list of vowels and a list of words that are difficult to pronounce and write.

3   See introduction to volume 2 of Cuervo's *Obras Completas* [Complete Works] published by the Instituto Caro y Cuervo in 1987 for the dates and content of the different editions.

4   His work notoriously consists of reeditions and revisions of his own work, constantly revising every detail. For example, although Cuervo died in 1911, he managed to partially follow the editorial process of the final, 1914 edition of his *Apuntaciones*, which contains a greatly expanded prologue. The transformations in his understandings of language change, etymology and lexicography become evident in the differences between the first and the sixth prologue. By the sixth edition, the *Apuntaciones* had received the critique of recognized linguists of the period such as Reinhardt Dozy, Juan Eugenio Hartzenbuch, and August Friederich Pott. The sixth edition addressed some of these conversations and their critiques. During this period, he also had engaged in a highly contested polemic with an unnamed journalist in Argentina, in which he had to carefully and with scientific detail defend his ideas about the impossibility of maintaining the unity of the Spanish language between Spain and the Americas (see Del Valle 2002 for a detailed analysis of this polemic). As a result, the prologue to the sixth edition of the *Apuntaciones* also appeared as an independent text entitled *El castellano en América* (Castilian in America) and again as *Disquisiciones sobre filología castellana* [Disquisitions on Castilian Philology], all of which incorporate detailed attention to his linguistic precepts.

5   I will not dwell here on the differential implications of this when such truth value is mobilized today either by social movements or by the nation-state. What is

significant to point out is that, even within a neoliberal economy, the differential politics of sovereignty generated by the politics of the prior for indigenous peoples (or peoples claiming the prior as a means to political recognition) and those who govern have become even more strategic. This has generated a need to critically rethink their place within the contemporary changing relations between indigenous geontologies, neoliberal economies, sovereignty, and the nation-state (see Povinelli 2011; Leach 2007).

6   This is how he describes his work in his dictionary: Fijado el carácter gramatical primario de cada voz, se da al principio, cuando parece necesario, una idea del desenvolvimiento de las acepciones; explícanse luego éstas por su orden, así como las construcciones á que se prestan, y compruébanse y esclárense con ejemplos, acompañados de la indicación precisa de la edición de que se toman, que es á menudo la Biblioteca de autores españoles de Rivadeneyra, no tanto en razón de su mérito (que en ocasiones es bien escaso), como en atención á lo accesible que es a toda suerte de lectores; algunas veces se comentan estos ejemplos ó se les agregan las indicaciones filológicas bastantes á asegurar la interpretación. Vienen en seguida las autoridades del período o anteclásico dispuestas aproximativamente en orden cronológico ascendente, y en seguida los testimonios sacados de documentos latinos ó cuasilatinos redactados en España antes de ser el castellano idioma oficial. Cierra el cuadro la etimología ó su discusión cuando no es clara. Además, si las palabras dan ocasión a ello, se anotan los accidentes morfológicos, prosódicos y ortográficos, y en los artículos de verbos, cuando son largos ó complicados, va al fin un índice de las construcciones (1886, liii).

7   As far as I understand the system, these are arbitrary assignments.

8   For further elaboration on this transformation, see Chua 1999 and Moreno 2004, among others.

9   For the complex rearrangements between instrumental and vocal music that this implied, see Chua 1999.

Epilogue: The Oral in the Aural

1   Faudree sees music and language as constituting a "communicative whole." Even though many aspects of both overlap, other dimensions also set them apart as distinct phenomena, even if one can be metaphorized easily as the other. This generates the question whether music or language always or primarily communicate. On the use of metaphor as a ground for thinking and a means to understand the relation between nature and culture or the ontological and epistemological across different cultures, see Wagner [1975] 1981.

REFERENCES

Abadía Morales, Guillermo. 1985. "Un hallazgo literario." In Jorge Isaacs, *Canciones y coplas populares*. Bogotá: Procultura.

Abbate, Carolyn. 1991. *Unsung Voices: Opera and Musical Narrative in the Nineteenth Century*. Princeton, NJ: Princeton University Press.

Adam, Lucien, and Victor Henry. 1880. *Arte y vocabulario de la lengua chiquita: Con algunos textos traducidos y explicados, compuestos sobre manuscritos inéditos del XVIII* siglo. Paris: Maisonneuve.

Adler, Guido. [1885] 1981. "Guido Adler's The Scope, Method and Aim of Musicology: An English Translation with a Historico-Analytical Commentary by Erica Mugglestone." *Yearbook for Traditional Music* 13: 1–21.

Almario G., Oscar. 2005. *La invención del suroccidente colombiano*. Medellín: Universidad Pontífica Bolivariana.

Almario G., Oscar. 2007. "Los paisajes ocultos y la invisibilidad de los 'otros' en Jorge Isaacs." In *Jorge Isaacs, el creador en todas sus facetas*, edited by Darío Henao Restrepo, 213–30. Cali: Universidad del Valle.

Alter, Stephen G. 1999. *Darwinism and the Linguistic Image: Language, Race, and Natural Theology in the Nineteenth Century*. Baltimore: Johns Hopkins University Press.

Ames, Eric. 2003. "The Sound of Evolution." *modernism/modernity* 10 (2): 297–325.

Anidjar, Gil. 2002. *"Our Place in al-Andalus": Kabbalah, Philosophy, Literature in Arab Jewish Letters*. Palo Alto, CA: Stanford University Press.

Anidjar, Gil. 2003. *The Jew, the Arab: A History of the Enemy*. Palo Alto, CA: Stanford University Press.

Anidjar, Gil. 2008. *Semites: Race, Religion, Literature*. Palo Alto, CA: Stanford University Press.

Arango, Rodolfo. 2002. "La construcción de la nacionalidad." In *Miguel Antonio Caro y la cultura de su época*, edited by Rubén Sierra Mejía, 125–53. Bogotá: Universidad Nacional de Colombia.

Arendt, Hannah. [1958] 1998. *The Human Condition*. Chicago: University of Chicago Press.

Arias, Julio, and Eduardo Restrepo. 2010. "Historizando raza: Propuestas conceptuales y metodológicas." *Emancipación y crítica* 3: 45–64.

Arias de Greiff, Jorge. 1969. "El diario inédito de Humboldt." *Revista de la Academia Colombiana de Ciencias Exactas y Naturaleza* 13 (51): 393–402.

Arocha, Jaime. 2004. *Utopía para los excluidos: El multiculturalismo en Africa y América Latina*. Bogotá: Facultad de Ciencias Humanas, Colecció CES, Universidad Nacional de Colombia.

Asad, Talal. 1993. *Genealogies of Religion: Discipline and Reasons of Power in Christianity and Islam*. Baltimore: Johns Hopkins University Press.

Asad, Talal. 2003. *Formations of the Secular, Christianity, Islam, Modernity*. Palo Alto, CA: Stanford University Press.

Augoyard, Jean-Francois, and Henry Torgue, eds. 2005. *Sonic Experience: A Guide to Everyday Sounds*. Montreal: McGill-Queen's University Press.

Avelar, Idelber. 2004. *The Letter of Violence: Essays on Narrative, Ethics, and Politics*. New York: Palgrave Macmillan.

Avelar, Idelber. 2013. "Amerindian Perspectivism and Non-human Rights." *alter/nativas* 1: 2–21.

Baena, Juan Alfonso de. 1851. *El cancionero de Juan Alfonso de Baena (siglo XV), ahora por primera vez dado á luz, con notas y comentarios*. Madrid: Imprenta de la Publicidad.

Baker, Geoffrey. 2008. *Imposing Harmony: Music and Society in Colonial Cuzco*. Durham, NC: Duke University Press.

Basso, Ellen B. 1985. *A Musical View of the Universe: Kalapalo Myth and Ritual Performances*. Philadelphia: University of Pennsylvania Press.

Bauman, Richard. 1984. *Verbal Art as Performance*. Prospect Heights, IL: Waveland Press.

Bauman, Richard. 2004. Introduction, 1–14. *A World of Others' Words: Cross-cultural Perspectives on Intertextuality*. Malden, MA: Blackwell Publishers.

Bauman, Richard, and Charles L. Briggs. 2003. *Voices of Modernity: Language Ideologies and the Politics of Inequality*. Cambridge: Cambridge University Press.

Bello, Andrés. 1905. *Gramática de la lengua castellana destinada al uso de los americanos*, edited by Rufino José Cuervo. Paris: R. Roger and F. Chernoviz.

Beltrán Pepió, Vicenç. 1998. "The Typology and Genesis of the Cancioneros: Compiling the Materials." In *Poetry at Court in Trastamaran Spain: From the Cancionero de Baena to the Cancionero General*, edited by E. Michael Gerli and Julian Weiss, 19–46. Tempe: University of Arizona Press.

Benes, Tuska. 2004. "Comparative Linguistics as Ethnology: In Search of Indo-Germans in Central Asia, 1770–1830." *Comparative Studies of South Asia, Africa and the Middle East* 24 (2): 117–32.

Benes, Tuska. 2008. *In Babel's Shadow: Language, Philology and the Nation in Nineteenth-Century Germany*. Detroit: Wayne State University Press.

Benjamin, Walter. [1955] 1968. "Theses on the Philosophy of History." In *Illuminations, Essays and Reflections*, edited by Hannah Arendt, 253–64. New York: Schocken Books.

Benjamin, Walter. [1955] 1968a. "What Is Epic Theater?" In *Illuminations, Essays and Reflections*, edited by Hannah Arendt, 147–54. New York: Schocken Books.

Bermúdez, Egberto, with Ellie Anne Duque. 2000. *Historia de la música en Santa Fé y Bogotá 1538–1938*. Bogotá: Fundación de Música.

Bernand, Carmen, and Serge Gruzinski. 1988. *De l'idolâtrie: Une archéologie des sciences religieuses*. Paris: Editions du Seuil.

Bird-David, Nurit. 1999. "'Animism' Revisited: Personhood, Environment and Relational Epistemology." *Current Anthropology, Special Issue, Culture: A Second Chance?* 40 (s1), s67–s91.

Blacking, John. 1973. *How Musical Is Man?* Seattle: University of Washington Press.

Bloechl, Olivia. 2008. *Native American Song at the Frontiers of Early Modern Music.* Cambridge: Cambridge University Press.

Borda, José Joaquín. 1866. "Seis horas en un Champán: Museo de cuadros de costumbres II." Bogotá: D.F. Accessed May 2009. http://www.lablaa.org/blaavirtual/literatura/cosii/indice.htm.

Borja Gómez, Jaime Humberto. 1996. *Inquisición, muerte y sexualidad en el Nuevo Reino de Granada.* Santafé de Bogotá: Editorial Ariel.

Botero, Clara Isabel. 2002. "Ezequiel Uricoechea en Europa: Del naturalismo a la filología." *Boletín Cultural y Bibliográfico* 39 (59): 3–27.

Brady, Erika. 1999. *Spiral Way: How the Phonograph Changed Ethnography.* Jackson: University Press of Mississippi.

Breva-Claramonte, Manuel. 2001. "Data Collection and Data Analisis in Lorenzo Hervás, Laying the Ground for Modern Linguistic Typology." In *History of Linguistics in Spain = Historia de la lingüistica en España, volume II,* edited by E. F. K. Koerner and Hans-Josef Niederehe, 265–80. Series 3, Studies in the History of the Language Sciences, 100. Amsterdam: J. Benjamins.

Bryant, Levi, Nick Srnicek, and Graham Harman. 2010. *The Speculative Turn: Continental Materialism and Realism.* Prahranm Vic, Australia: re.press.

Bunzl, Matti. 1996. "Franz Boas and the Humboldtian Tradition: From Volkgeist and Nationalcharakter to an Anthropological Notion of Culture." In *Volkgeist as Method and Ethic: Essays on Boasian Ethnography and the German Anthropological Tradition,* edited by George Stocking, 17–78. Madison: University of Wisconsin Press.

Bunzl, Matti, and H. Glenn Penny. 2010. "Introduction: Rethinking German Anthropology, Colonialism and Race." In *Wordly Provincialism, German Anthropology in the Age of Empire,* edited by H. Glenn Penny and Matti Bunzl, 1–30. Ann Arbor: University of Michigan Press.

Cañizares-Esguerra, Jorge. 2001. *How to Write the History of the New World: Histories, Epistemologies, and Identities in the Eighteenth-Century Atlantic World.* Palo Alto, CA: Stanford University Press.

Cañizares-Esguerra, Jorge. 2006. *Nature, Empire, and Nation: Explorations of the History of Science in the Iberian World.* Palo Alto, CA: Stanford University Press.

Cañizares-Esguerra, Jorge. 2006a. *Puritan Conquistadors: Iberianizing the Atlantic, 1550–1700.* Palo Alto, CA: Stanford University Press.

Cañizares-Esguerra, Jorge. 2006b. "Creole Colonial Spanish America." In *Creolization: History, Ethnography, Theory,* edited by Charles Stewart, 26–45. Walnut Creek, CA: Left Coast Press.

Caraballo, Vicente. 1943. *El Negro Obeso (Apuntes Biográficos) y Escritos Varios.* Bogotá: Editorial ABC.

Caro, Miguel Antonio. [1867] 1980. "Reacción ortográfica." In *Miguel Antonio Caro, Obras, Estudios lingüísticos, gramaticales y filológicos,* vol. 3, edited by Carlos Valderrama Andrade, 354–73. Bogotá: Instituto Caro y Cuervo, Clásicos colombianos VIII.

Caro, Miguel Antonio. [1867] 1980a. "Ortografia castellana." In *Miguel Antonio Caro, Obras, Estudios lingüísticos, gramaticales y filológicos,* vol. 3, edited by Carlos Valderrama Andrade, 374–80. Bogotá: Instituto Caro y Cuervo, Clásicos colombianos VIII.

Caro, Miguel Antonio. [1881] 1980a. "Del uso en sus relaciones con el lenguaje." In *Miguel Antonio Caro, Obras, Estudios linguísticos, gramaticales y filológicos*, vol. 3, edited by Carlos Valderrama Andrade, 3–80. Bogotá: Instituto Caro y Cuervo, Clásicos colombianos VIII.

Caro, Miguel Antonio. [1881] 1980. "Manual de elocución." In *Miguel Antonio Caro, Obras, Estudios lingüísticos, gramaticales y filológicos*, vol. 3, edited by Carlos Valderrama Andrade, 429–66. Bogotá: Instituto Caro y Cuervo.

Caro, Miguel Antonio. [1886] 1980. "El Darwinismo y las misiones." In *Miguel Antonio Caro, Obras Completas, Filosofía, Religión, Pedagogía*, vol. 1, edited by Carlos Valderrama Andrade, 1049–107. Bogotá: Instituto Caro y Cuervo.

Caro, Miguel Antonio. [1886] 1980a. "La Religión de la Nación." In *Miguel Antonio Caro, Obras, Filosofía, Religión, Pedagogía*, vol. 1, edited by Carlos Valderrama Andrade, 1035–48. Bogotá: Instituto Caro y Cuervo.

Caro, Miguel Antonio. [1871] 1980. "Autoridad es Razón." In *Miguel Antonio Caro, Obras, Filosofía, Religión, Pedagogía*, vol. 1, edited by Carlos Valderrama Andrade, 562–67. Bogotá: Instituto Caro y Cuervo.

Caro, Miguel Antonio. [1884] 1980. "Fonética y ortografía." In *Miguel Antonio Caro, Obras, Estudios lingüísticos, gramaticales y filológicos*, vol. 3, edited by Carlos Valderrama Andrade, 329–471. Bogotá: Instituto Caro y Cuervo.

Caro, Miguel Antonio. [1928] 1980. "Papeles sueltos." In *Miguel Antonio Caro, Obras, Estudios lingüísticos, gramaticales y filológicos*, vol. 3, edited by Carlos Valderrama Andrade, 329–53. Bogotá: Instituto Caro y Cuervo, Clásicos colombianos VIII.

Carter, Paul. 2001. "Ambiguous Traces: Mishearing and Auditory Space." In *Australian Sound Design Project*. Accessed September 2012. http://www.sounddesign.unimelb.edu.au/site/papers/mishearing.html.

Casanova, José. 1994. *Public Religions in the Modern World*. Chicago: University of Chicago Press.

Castro-Gómez, Santiago. 2004. *La hybris del punto cero: Ciencia, raza e ilustración en la Nueva Granada, 1750–1816*. Bogotá: Editorial Javeriana.

Castro-Gómez, Santiago, and Ramón Grosfoguel, eds. 2007. "Prólogo: Giro decolonial, teoría crítica y pénsamiento heterárquico," 9–24. *El giro decolonial: Reflexiones para una diversidad epistémica más allá del capitalismo global*. Bogotá: Pontificia Universidad Javeriana. Universidad Central, Siglo del Hombre Editores.

Cavarero, Adriana. 2005. *For More than One Voice: Toward a Philosophy of Vocal Expression*. Palo Alto, CA: Stanford University Press.

Ceballos Gómez, Diana Luz. 1994. *Hechicería, brujería e Inquisición en el Nuevo Reino de Granada: Un duelo de imaginarios*. Medellín: Editorial Universidad Nacional.

Ceballos Gómez, Diana Luz. 2001. "Prácticas mágicas y grupos sociales en el Nuevo Reino de Granada." In *Historia Crítica 18* (1), 1–20.

Ceballos Gómez, Diana Luz. 2005. "Iconografía y guerras civiles en la Colombia del siglo XIX: Una mirada a la representación." In *Ganarse el cierlo defendiendo la religión, Guerras civiles en Colombia 1840–1902*, edited by Grupo de Investigación Religión, Cultura y Sociedad, Universidad Nacional, 157–210. Bogotá: Universidad Nacional de Colombia.

Chakrabarty, Dipesh. 2000. *Provincializing Europe: Postcolonial Thought and Historical Difference*. Princeton, NJ: Princeton University Press.

Chakrabarty, Dipesh. 2009. "The Climate of History: Four Theses." *Critical Inquiry* 35: 197–222.

Chua, Daniel K. L. 1999. *Absolute Music and the Construction of Meaning*. Cambridge: Cambridge University Press.

Clark, Suzannah, and Alexander Rehding, eds. 2001. "Introduction." *Music Theory and the Natural Order from the Renaissance to the Early Twentieth Century*. Cambridge: Cambridge University Press.

Coccia, Emanuele. 2011. *La vida sensible*. Translated by Maria Teresa d'Meza. Buenos Aires: Editorial Marea.

Cochrane, Charles Stuart. 1825. *Journal of a Residence and Travels in Colombia during the Years 1823 and 1824*. Vols. 1 and 2. London: Printed for Henry Colburn.

Colmenares, Germán. [1973] 1991. *Historia económica y social de Colombia 1537–1719*. Cali: Universidad del Valle.

Colmenares, Germán. 1976. *Cali: Terratenientes, mineros y comerciantes, siglo XVIII*. Cali: Universidad del Valle.

Connor, Steven. 2000. *Dumbstruck: A Cultural History of Ventriloquism*. Oxford: Oxford University Press.

Connor, Steven. 2004. "Sound and the Self." In *Hearing History, a Reader*, edited by Mark M. Smith, 54–68. Athens: University of Georgia Press.

Connor, Steven. 2006. "Phonophobia: The Dumb Devil of Stammering." A talk given, as "Giving Out Voice," at the Giving Voice conference, University of Aberystwyth, April 8, 2006. http://www.stevenconnor.com/phonophobia/.

Connor, Steven. 2009. "Earslips: Of Mishearings and Mondegreens." A talk given at the conference Listening In, Feeding Back, Columbia University, February 14, 2009. http://www.stevenconnor.com/earslips/.

Corbin, Alain. 1998. *Village Bells: Sound and Meaning in the Nineteenth-Century French Countryside*. New York: Columbia University Press.

Cournut, Jean. 1974. "De l'écriture à l'inscription ou *le scribe de l'inconscient*." *Revue Française de Psychanalyse* 38: 57–74.

Crary, Jonathan. 1992. *Techniques of the Observer: On Vision and Modernity in the Nineteenth Century*. Cambridge, MA: MIT Press.

Cristina, María Teresa, ed. 2005. "Introducción." *Isaacs: Obras Completas. Vol 1. María*. Bogotá: Universidad Externado de Colombia. Critical edition, prologue, introduction and notes by María Teresa Cristina, xxix–xlviii. Cali: Universidad del Valle.

Cristina, María Teresa, ed. 2006. "Introducción." *Isaacs: Obras Completas. Vol 2. Poesía*. Bogotá: Universidad Externado de Colombia. Critical edition, introduction, prologue and notes by María Teresa Cristina, xxiii–lxxxv. Cali: Universidad del Valle.

Crosby, Alfred W. 1986. *Ecological Imperialism: The Biological Expansion of Europe*. Cambridge: Cambridge University Press.

Cuervo, Rufino José. [1867; 1914 (6th ed.)] 1987. *Apuntaciones críticas sobre el lenguaje bogotano*. Bogotá: Instituto Caro y Cuervo.

Cuervo, Rufino José. 1886. *Diccionario de construcción y régimen de la lengua castellana*, vol. 1. Paris: A. Roger and F. Chernoviz.

Cuervo, Rufino Jose. 1905. *Prólogo y anotaciones a la gramática de la lengua castellana de Andrés Bello: Novena edición hecha sobre la última del autor con extensas notas y un copioso índice alfabético*. Paris: A. Roger and F. Chernoviz.

Cuervo, Rufino José. 1935. *El castellano en América*. Colección Biblioteca Aldeana de Colombia. No. 2. Bogotá: Editorial Minerva.

Cuervo, Rufino José. 1976. *Epistolario de Rufino José Cuervo con filólogos de Alemania, Austria y Suiza*, Vol. 2. Edited by Günther Schutz. Bogotá: Instituto Caro y Cuervo.

Cuervo, Rufino José. 1978. *Epistolario de Rufino José Cuervo con Miguel Antonio Caro*. Introducción y notas sur Mario Germán Romero. Bogotá: Instituto Caro y Cuervo.

Cunin, Elisabeth. 2002. "Asimilación, multiculturalismo y mestizaje: Formas y transformaciones de la relación con el otro en Cartagena." In *Afrodescendientes en las américas, trayectorias sociales e identitarias: 150 años de la abolición de la esclavitud en Colombia*, edited by Claudia Mosquera, Mauricio Pardo, and Odile Hoffman, 279–94. Bogotá: Universidad Nacional de Colombia, Instituto Colombiano de Antropología e Historia, Institut de Recherche pour le Développement, Instituto Latinoamericano de Servicios Legales Alternativos.

Cunin, Elisabeth. 2002a. "Chicago bajo el trópico o las virtudes heurísticas del mestizaje." *Revista colombiana de antropología* 38: 11–44.

Cunin, Elisabeth. 2003. *Identidades a Flor de Piel, lo "negro" entre apariencias y presencias: Categorías raciales y mestizaje en Cartagena*. Bogotá: Instituto Colombiano de Antropología e Historia, Universidad de los Andes, Instituto Francés de Estudios Andinos, Observatorio del Caribe Colombiano.

Dalhaus, Carl. 1982. *Esthetics of Music*. Cambridge: Cambridge University Press.

Davies, Anna Morpurgo. 1998. "Nineteenth Century Linguistics." In *History of Linguistics*, vol. 4, edited by Giulio C. Lepschy. London: Longman.

De la Cadena, Marisol. 2000. *Indigenous Mestizos: The Politics of Race and Culture in Cuzco, Perú, 1919–1991*. Durham, NC: Duke University Press.

De la Campa, Román. 1999. *Latin Americanism*. Minneapolis: University of Minnesota Press.

De Landa, Manuel. 1997. *A Thousand Years of Nonlinear History*. New York: Swerve Editions.

De Landa, Manuel. 2006. *A New Philosophy of Society: Assemblage Theory and Social Complexity*. London: Continuum.

Deas, Malcolm. [1992] 2006. "Miguel Antonio Caro y amigos: Gramática y poder en Colombia." *Del poder y la gramática y otros ensayos sobre historia, política y literatura colombianas*. Bogotá: Taurus, 27–62.

Del Sarto, Ana, Alicia Ríos, and Abril Trigo. 2004. *The Latin American Cultural Studies Reader*. Durham, NC: Duke University Press.

Del Valle, José. 2002. "Historical Linguistics and Cultural History: The Polemic between Rufino José Cuervo and Juan Valera." In *The Battle over Spanish between 1800 and 2000*, edited by José del Valle and Luis Gabriel Stheeman, 64–77. New York: Routledge.

Del Valle, José. 2002a. "Lenguas imaginadas: Menéndez Pidal, la lingüística hispánica y la configuración del estándar." In *Estudios de lingüística del español*, 16. Accessed January 2013. http://elies.rediris.es/elies16/Valle.html#n1.

Deleuze, Gilles. 1969. *Logique du Sens*. Paris: Les Éditions de Minuit.

Deleuze, Gilles, and Félix Guattari. [1987] 2011. *A Thousand Plateaus: Capitalism and Schizophrenia*. Translation and foreword by Brian Massumi. Minneapolis: University of Minnesota Press.

Derrida, Jacques. [1974] 1997. *Of Grammatology*. Baltimore: Johns Hopkins University Press.

Derrida, Jacques. 2002. "Des tours de Babel." In *Acts of Religion: Jacques Derrida*, edited by Gil Anidjar, 102–34. New York: Routledge.

Derrida, Jacques. [1978] 2005. *Writing and Difference*. Translated by Alan Bass. Routledge: London.

Dijk, P. van, S.J. 1952. "Medieval Terminology and Methods of Psalm Singing." *Musica Disciplina* 6 (1/3): 7–26.

Dolar, Mladen. 2006. *A Voice and Nothing More*. Cambridge, MA: MIT Press.

Drobnick, Jim, ed. 2004. "Listening Awry." *Aural Cultures*, 6–18. Toronto: YYZ Books.

Duffey, Frank M. 1956. *The Early Cuadro de Costumbres in Colombia*. Chapel Hill: University of North Carolina Press.

Ellis, Alexander J. 1885. "On the Musical Scales of Various Nations." *Journal of the Society of the Arts* 33 (1688), 485–525.

Erlmann, Veit. 1999. *Music, Modernity and the Global Imagination: South Africa and the West*. New York: Oxford University Press.

Erlmann, Veit, ed. 2004. "But What of the Ethnographic Ear? Anthropology, Sound and the Senses." *Hearing Cultures, Essays on Sound, Listening, and Modernity*, 1–20. Oxford: Berg.

Erlmann, Veit. 2010. *Reason and Resonance: A History of Modern Aurality*. New York: Zone Books.

Escobar, Arturo. 2008. *Territories of Difference: Place, Movements, Life, Redes*. Durham, NC: Duke University Press.

Escobar, Ticio. [1993] 2012. *La belleza de los otros: Arte indígena del Paraguay*. Asunción, Paraguay: Servilibro.

Esposito, Roberto. 2008. *Términos de la política*. Barcelona: Herder.

Esposito, Roberto. [1998] 2009. *Comunidad, inmunidad y biopolítica*. Translated by Alicia García Ruiz. Barcelona: Herder.

Esposito, Roberto. 2009a. *Tercera Persona: Política de la vida y filosofía de lo impersonal*. Buenos Aires: Amorrortu.

Fallón, Diego. 1869. *Nuevo sistema de escritura musical*. Bogotá: Imprenta Metropolitana.

Fallón, Diego. 1885. *Arte de leer, escribir y dictar música, Sistema Alfabético*. Bogotá: Imprenta Musical.

Farinelli, Franco. 2009. *La crisi della ragione cartografica: Introduzione alla geografia della globalità*. Turin: Einaudi.

Faudree, Paja. 2012. "Music, Language and Texts: Sound and Semiotic Ethnography." In *Annual Review of Anthropology* 41: 519–36.

Feld, Steven. 1996. "Waterfalls of Song: An Acoustemology of Place Resounding in Bosavi, Papua New Guinea." In *Senses of Place*, edited by Steven Feld and Keith H. Basso, 91–136. Santa Fe, NM: School of American Research Press.

Feld, Steven. 2012. *Jazz Cosmopolitanism in Accra: A Memoir of Five Musical Years in Ghana*. Durham, NC: Duke University Press.

Feld, Steven, and Aaron Fox. 1994. "Music and Language." *Annual Review of Anthropology* 23: 25–53.

Feld, Steven, Aaron A. Fox, Thomas Porcello, and David Samuels. [2004] 2006. "Vocal Anthropology: From the Music of Language to the Language of Song." In *A Companion to Linguistic Anthropology*, edited by Alessandro Duranti, 321–46. Malden, MA: Blackwell.

Fessel, Pablo. 2000. Unpublished article. "From Stylistic Categories to the Concept of Texture: Changes in the Representation of Simultaneity in the Music Thinking of the Early Twentieth Century."

Fox, Aaron. 2004. *Real Country: Music and Language in Working-class Culture*. Durham, NC: Duke University Press.

Franco, Jean. 2002. *The Decline and Fall of the Lettered City: Latin America in the Cold War*. Cambridge, MA: Harvard University Press.

Funes, Leonardo. 2003. "La apuesta por la historia de los habitantes de la Tierra Media." In *Propuestas teórico-metodológicas para el estudio de la literatura hispánica medieval*, edited by Liliana Von der Walde Moheno, 15–34. Mexico City: Instituto de Investigaciones Filológicas, Universidad Nacional Autónoma de México, Universidad Autónoma Metropolitana, Difusión Cultural.

Gaitán Bohórquez, Julio César. 2012. "Domesticar la barbarie y dulcificar las costumbres: homogenización reductora y proyecto jurídico nacional." In *Escuela Intercultural de Diplomacia Indígena: Memoria, derecho y participación: La experiencia del pueblo arhuaco, nabusímake, Sierra Nevada de Santa Marta*, edited by Ana Rodríguez, Pedro Rojas, Ángela Santamaría, 205–24. Bogotá: Universidad del Rosario.

García Bernal, Ricardo. 2007. *Juan Bernardo Elbers: Del Rhin al Magdalena*. Bogotá: R. García Bernal.

Garrido, Margarita. 2007. "Libres de todos los colores en Nueva Granada: Identidad y obediencia antes de la independencia." In *Cultura política en los Andes*, coordinated by Nils Jacobsen and Cristóbal Aljovín, 245–66. Lima: Universidad Nacional Mayor de San Marcos.

Gitelman, Lisa. 1999. *Scripts, Grooves and Writing Machines: Representing Technology in the Edison Era*. Palo Alto, CA: Stanford University Press.

Gitelman, Lisa. 2006. *Always Already New: Media, History and the Data of Culture*. Cambridge, MA: MIT Press.

Goehr, Lydia. 1992. *The Imaginary Museum of Musical Works: An Essay in the Philosophy of Music*. Oxford: Oxford University Press.

Goldstein, E. Bruce, ed. 2001. *Blackwell Handbook of Perception*. Malden, MA: Blackwell.

Gómez-Bravo, Ana M. 1999. "Retórica y poética en la evolución de los géneros cuatrocentista." *Rhetorica: A Journal of the History of Rhetoric* 17 (2): 137–75.

González Echevarría, Roberto. 1990. *Myth and Archive: A Theory of Latin American Narrative*. Cambridge: Cambridge University Press.

González González, Fernán E. 1997. *Poderes enfrentados: Iglesia y estado en Colombia*. Bogotá: CINEP.

González Lineros, Narciso. [1877] 1885. "Nuevo Sistema Musical." In Fallón, Diego. 1885. *Arte de leer, escribir y dictar música, Sistema Alfabético*, 8–9. Bogotá: Imprenta Musical.

Goodman, Steve. 2010. *Sonic Warfare: Sound, Affect, and the Ecology of Fear.* Cambridge, MA: MIT Press.

Gosselman, Carl August. [1830] 1981. *Viaje por Colombia: 1825 y 1826.* Spanish version by Ann Christien Pereira. Bogotá: Ediciones del Banco de la República, 1981. Document digitalized by the Virtual Library of the Banco de la República in 2005. Accessed February and March 2008. *http://www.lablaa.org/blaavirtual/historia/viajes/indice.htm.*

Gray, Ellen. 2013. *Fado Resounding: Affective Politics and Urban Life.* Durham, NC: Duke University Press.

Green, Burdette, and David Butler. 2002. "From Acoustics to Tonpsychologie." In *The Cambridge History of Western Music Theory,* edited by Thomas Christensen, 246–71. Cambridge: Cambridge University Press.

Grosz, Elizabeth. 2004. *The Nick of Time: Politics, Evolution, and the Untimely.* Durham, NC: Duke University Press.

Grosz, Elizabeth. 2011. *Becoming Undone: Darwinian Reflections on Life, Politics, and Art.* Durham, NC: Duke University Press.

Guss, David M. 1989. *To Weave and Sing: Art, Symbol, and Narrative in the South American Rainforest.* Berkeley: University of California Press.

Hanks, William F. 2000. "Authenticity and Ambivalence in the Text: A Colonial Maya Text." *Intertexts: Writings on Language, Utterance, and Context,* 103–32. Lanham, MD: Rowman and Littlefield.

Hanks, William F. 2010. *Converting Words: Maya in the Age of the Cross.* Berkeley: University of California Press.

Hanslick, Eduard. [1885] 1891. *The Beautiful in Music: A Contribution to the Revisal of Musical Aesthetics.* 7th edition translated by Gustav Cohen. London: Novello.

Helg, Aline. 2004. *Liberty and Equality in Caribbean Colombia, 1770–1835.* Chapel Hill: University of North Carolina Press.

Helmrich, Stefan. 2007. "An Anthropologist under Water: Immersive Soundscapes, Submarine Cyborgs, and Transductive Ethnography." *American Anthropologist* 34 (4): 621–41.

Hernández de Alba, Gregorio. 1968. *Ezequiel Uricochea: Noticia bibliográfica y homenaje en la ciudad de Bruselas.* Bogotá: Instituto Caro y Cuervo.

Higashi, Alejandro. 2003. "Edad media y geneología: El caso de las etiquetas de género." In *Propuestas teórico-metodológicas para el estudio de la literatura hispánica medieval,* edited by Liliana Von der Walde Moheno, 35–73. Mexico City: Instituto de Investigaciones Filológicas, Universidad Nacional Autónoma de México, Universidad Autónoma Metropolitana, Difusión Cultural.

Hill, Jonathan D. 1993. *Keepers of the Sacred Chants: The Poetics of Ritual Power in an Amazonian Society.* Tucson: University of Arizona Press.

Hill, Jonathan D. 2009. *Made-from-Bone: Trickster Myths, Music, and History from the Amazon.* Urbana: University of Illinois Press.

Hill, Jonathan D., and Jean-Pierre Chaumeil. 2011. *Burst of Breath: Indigenous Ritual Wind Instruments in Lowland South America.* Lincoln: University of Nebraska Press.

Hilmes, Michelle. 2005. "Is There a Field Called Sound Culture Studies? And Does It Matter?" *American Quarterly* 57 (1): 249–59.

Hirschkind, Charles. 2006. *The Ethical Soundscape: Cassette Sermons and Islamic Counterpublics.* New York: Columbia University Press.

Holbraad, Martin. 2012. *Truth in Motion: The Recursive Anthropology of Cuban Divination*. Chicago: University of Chicago Press.

Holton, Isaac F. 1857. *New Granada: Twenty Months in the Andes*. New York: Harper and Brothers.

Howes, David. 2004. *Sensual Relations: Engaging the Senses in Culture and Social Theory*. Ann Arbor: University of Michigan Press.

Humboldt, Alexander von. 1801. *Alexander von Humboldt en Colombia. Extractos de sus diarios*. Biblioteca Digial Andina. Accessed May 2009. http://www.comunidadandina.org/bda/docs/CO-CA-0004.pdf.

Humboldt, Alexander von. 1850. *Views of Nature: Or Contemplations on the Sublime Phenomena of Creation, with Scientific Illustrations*. Translated by E. C. Otté and H. G. Bohn. London: Henry G. Bohn.

Humboldt, Alexander von. [1853] 1971. *Personal Narrative of Travels to the Equinoctial Regions of America, during the Years 1799–1804, by Alexander von Humboldt and Aimé Bonpland*, vols. 1–3. New York: B. Blom.

Humboldt, Alexander von. 1858. *Cosmos: A Sketch of a Physical Description of the Universe*, vol. 1. Translated by E. C. Otté. New York: Harper and Brothers.

Humboldt, Alexander von. 1980. *Cartas americanas*. Venezuela: Biblioteca Ayacucho.

Ingold, Tim. 2011. *Being Alive: Essays on Movement, Knowledge and Description*. London: Routledge.

Irigaray, Luce. 1996. *I Love to You: Sketch for a Felicity within History*. New York: Routledge.

Isaacs, Jorge. [1884] 1959. *Estudio sobre las tribus indígenas del Magdalena*. Bogotá: Ministerio de Educación Nacional.

Isaacs, Jorge. 1985. *Canciones y coplas populares*, edited by Guillermo Abadía Morales. Bogotá: Procultura.

Isaacs, Jorge. [1867] 2005. *María*, vol. 1, Complete Works. Critical edition, prologue, and notes by María Teresa Cristina. Bogotá: Universidad Externado de Colombia.

Isaacs, Jorge. 2006. *Poesía, canciones y coplas populares, traducciones*, vol. 2, edited by María Teresa Cristina. Bogotá: Universidad Externado de Colombia.

Jankowsky, Kurt R. 1972. *The Neo-Grammarians: A Reevaluation of Their Place in the Development of Linguistic Science*. The Hague: Mouton.

Jaramillo Uribe, Jaime. 1989. *Ensayos de historia social*. Bogotá: Tercer Mundo Editores.

Jáuregui, Carlos. 2007. "Candelario Obeso, la literatura 'afronacional' y los límites del espacio literario decimonónico." In *"Chambacú, la historia la escribes tú": Ensayos sobre cultura afrocolombiana*, edited by Lucía Ortiz, 47–68. Madrid: Iberoamericana—Vervuert.

Jay, Martin. 1993. *Downcast Eyes: The Denigration of Vision in Twentieth-Century French Thought*. Berkeley: University of California Press.

Johnson, James H. 1995. *Listening in Paris: A Cultural History*. Berkeley: University of California Press.

Keane, Webb. 1999. "Voice." *Journal of Linguistic Anthropology* 9 (1–2): 271–73.

Keane, Webb. 2011. "Indexing Voice: A Morality Tale." In *Journal of Linguistic Anthropology* 21 (2), 166–78.

La Barrière, Jean-Louis. 1993. "Aristote et le problème du langage animal." *Metis* 8 (2), 247–60.

La Barrière, Jean-Louis. 2008. "Le caractère musicl de la voix chez Aristote *Apostasis, Melos, Dialektos.*" In *Vocaularies de la Voix*, edited by Barbara Cassin and Danielle Cohen-Levinas, 11–28. Paris: L'Harmattan.

Lacoue-Labarthe, Philippe, and Jean-Luc Nancy. 1988. *The Literary Absolute: The Theory of Literature in German Romanticism.* Albany: State University of New York Press.

Lafuente y Alcántara, Emilio. 1865. *Cancionero popular, colección escogida de seguidillas y coplas.* Vols. 1 and 2. Madrid: Carlos Billy-Bailliere.

Landers, Jane. 2002. "Conspiradores esclavizados en Cartagena en el siglo XVIII." In *Afrodescendientes en las américas, trayectorias sociales e identitarias: 150 años de la abolición de la esclavitud en Colombia*, edited by Claudia Mosquera, Mauricio Pardo and Odile Hoffman, 181–94. Bogotá: Universidad Nacional de Colombia, Instituto Colombiano de Antropología e Historia, Institut de Recherche pour le Développement, Instituto Latinoamericano de Servicios Legales Alternativos.

Langebaek, Carl. 2009. *Los herederos del pasado: Indígenas y pensamiento en Colombia y Venezuela.* Bogotá: Universidad de los Andes.

Latour, Bruno. 1993. *We Have Never Been Modern.* Cambridge, MA: Harvard University Press.

La Tronkal. 2010. *Desenganche, visualidades y sonoridades otras.* Quito: Abilit.

Leach, James. 2007. "Creativity, Subjectivity, and the Dynamic of Possessive Individualism." In *Creativity and Cultural Improvisation*, edited by Tim Ingold and E. Hallam, 99–116. ASA Monograph 43. Oxford: Berg.

Leal Buitrago, Francisco. 2000. "Vicisitudes de la profesionalización de las ciencias sociales en Colombia." In *Discurso y razón, una historia de las ciencias sociales en Colombia*, edited by Francisco Leal Buitrago and Germán Rey, 1–14. Bogotá: Ediciones Uniandes, Fundación Social, Tercer Mundo Editores.

Le Guin, Elisabeth. 2006. *Boccherini's Body: An Essay in Carnal Musicology.* Berkeley: University of California Press.

Le Moyne, Auguste. [1880] 1985. *Viaje y estancia en América del Sur, la Nueva Granada, Santiago de Cuba, Jamaica y el Istmo de Panamá.* Bogotá: Editorial Centro Instituto Gráfico, vol. 9, Biblioteca Popular de Cultura Colombiana.

León Gómez, Adolfo. 2002. "El estilo argumentativo de Miguel Antonio Caro." In *Miguel Antonio Caro y la cultura de su época*, edited by Rubén Sierra Mejía, 155–88. Bogotá: Universidad Nacional de Colombia.

Lienhard, Martin. [1990] 2011. *La voz y su huella.* Havana: Fondo Editorial Casa de las Américas.

Lévi-Strauss, Claude. [1955] 2012. *Tristes Tropiques.* London: Penguin Books.

Lévi-Strauss, Claude. [1964] 1983. *The Raw and the Cooked.* Chicago: University of Chicago Press.

Lévi-Strauss, Claude. 1966. *The Savage Mind.* Chicago: University of Chicago Press.

Lévi-Strauss, Claude. [1978] 1995. *Myth and Meaning.* New York: Schocken Books.

Lévi-Strauss, Claude. 1985. *The View from Afar.* New York: Basic Books.

Lomnitz, Claudio. 2001. *Deep Mexico, Silent Mexico: An Anthropology of Nationalism.* Minneapolis: University of Minnesota Press.

Lomnitz, Claudio. 2005. *Death and the Idea of Mexico.* New York: Zone Books.

Ludueña Romandini, Fabián. 2010. *La comunidad de los espectros. 1. Antropotecnia.* Buenos Aires: Miño y Dávila.

Luker, Morgan. 2007. "Tango Renovación: On the Uses of Music History in Post-Crisis Argentina." *Latin American Music Review* 28 (1): 68–93.

Maglia, Graciela. 2010. "Candelario Obeso a la luz del debate contemporáneo." In *Si yo fuera tambó: Poesía selecta de Candelario Obeso y Jorge Artel, edición crítica*, edited by Graciela Maglia, 17–26. Bogotá: Pontificia Universidad Javeriana y Universidad del Rosario.

Maniglier, Patrice. 2006. *La vie enigmatique des signes: Saussure et la naissance du structuralisme*. Paris: Scheer.

Maniglier, Patrice. 2010. *La perspective du Diable: Figurations de l'espace et philosophie de la Renaissance à Rosemary's Baby*. Nice, Villa Arson: Actes Sud.

Maniglier, Patrice, ed. 2011. *Le moment philosophique des années 1960 en France*. Paris: Presses Universitaries de France.

Manrique, Venancio G., and Candelario Obeso. 1884. *Lecciones prácticas de francés extractadas del curso completo de lengua francesa de T. Robertson*. Adaptación castellana. Bogotá: Librería de Cahves.

Marroquín, José Manuel. [1869] 1874. *Tratados de ortología y ortografía de la lengua castellana*. Bogotá: Imprenta de Gaitán.

Martín-Barbero, Jesús. [1987] 2001. *De los medios a las mediaciones: Comunicación, cultura y hegemonía*. Mexico City: Gustavo Gili.

Martín-Barbero, Jesús. 1998. "De la comunicación a la filosofía y viceversa: Nuevos mapas, nuevos retos." *Mapas Nocturnos Diálogos con la obra de Jesús Martín-Barbero*, 201–22. Bogotá: Fundación Universidad Central, Siglo XXI Editores.

Martín-Barbero, Jesús, and Hermann Herlinghaus. 2000. *Contemporaneidad latinoamericana y análisis cultural: Conversaciones al encuentro de Walter Benjamin*. Madrid: Iberoamericana.

Martínez, Frédéric. 2001. *El nacionalismo cosmopolita: La referencia europea en la construcción nacional en Colombia, 1845–1900*. Bogotá: Banco de la República/Instituto Francés de Estudios Andinos.

Masuzawa, Tomoko. 2005. *The Invention of World Religions, or, How European Universalism Was Preserved in the Language of Pluralism*. Chicago: University of Chicago Press.

Matory, James Lorand. 2005. *Black Atlantic Religion: Tradition, Transnationalism, and Matriarchy in the Afro-Brazilian Candomblé*. Princeton, NJ: Princeton University Press.

Maya Restrepo, Luz Adriana. 2005. *Brujería y reconstrucción de identidades entre los negros africanos y sus descendientes en la Nueva Granada, Siglo XVII*. Bogotá: Ministerio de Cultura.

Meintjes, Louise. 2005. "Shoot the Sergeant, Shatter the Mountain: The Production of Zulu Song and Dance in Post-Apartheid South Africa." *Ethnomusicology Forum* 13 (2).

Mejía Arango, Lázaro. 2007. *Los radicales: Políticas del radicalismo del siglo XIX*. Bogotá: Universidad Externado de Colombia.

Melo, Jorge Orlando. 2005. "La obra de Obeso tiene el homenaje de los lectores." In *Cantos populares de mi tierra*, 15–17. Bogotá: Alcaldía Mayor de Bogotá, Instituto Distrital de Cultura y Turismo—Observatorio de Cultura Urbana, Fundación Cultural y Ambiental Candelario Obeso.

Melo, Jorge Orlando. 2011. "Historia de la población y ocupación del territorio colombiano" (conferencia leída en 1990). In *Colombia es un tema.* http://www.jorgeorlando melo.com/histpobla.htm#_ftn1.

Menezes-Bastos, Rafael Jose de. 2007. "Música nas sociedades indígenas das terras baixas da América do Sul: Estado da arte." *Mana* 13 (2): 293–316.

Mignolo, Walter D. 1995. *The Darker Side of the Renaissance: Literacy, Territoriality, and Colonization.* Ann Arbor: University of Michigan Press.

Mignolo, Walter D. 2000. *Local Histories/Global Designs: Coloniality, Subaltern Knowledges, and Border Thinking.* Princeton, NJ: Princeton University Press.

Miller, Toby. 1993. *The Well-Tempered Self: Citizenship, Culture, and the Postmodern Subject.* Baltimore: Johns Hopkins University Press.

Miller, Toby, and George Yudice. 2002. *Cultural Policy.* London: Sage.

Miñana Blasco, Carlos. 1997. "Los caminos del bambuco en el siglo XIX." *A contratiempo* 9: 7–11.

Miñana Blasco, Carlos. 2000. "Entre el folklore y la etnomusicología: 60 años de estudios sobre la música popular tradicional en Colombia." *A contratiempo* 11: 36–49.

Minca, Claudio. 2007. "Humboldt's Compromise or the Forgotten Geographies of Landscape." *Progress in Human Geography* 31 (2): 179–93.

Montardo, Deise Lucy. 2009. *Através de Mbaraka: Musica, dança e xamanismo Guarani.* São Paulo: EDUSP.

Moraña, Mabel, Enrique Dussel, and Carlos A. Jáuregui. 2008. "Colonialism and Its Replicants." In *Coloniality at Large: Latin America and the Postcolonial Debate,* edited by Mabel Moraña, Enrique Dussel, and Carlos A. Jáuregui, 1–22. Durham, NC: Duke University Press.

Moreno, Jairo. 2004. *Musical Representations, Subjects, and Objects: The Construction of Musical Thought in Zarlino, Descartes, Rameau, and Weber.* Bloomington: Indiana University Press.

Morton, Timothy. 2007. *Ecology without Nature: Rethinking Environmental Aesthetics.* Cambridge, MA: Harvard University Press.

Mosquera, Claudia, Mauricio Pardo, and Odile Hoffman. 2002. "Las trayectorias sociales e identitarias de los afrodescendientes." In *Afrodescendientes en las Américas, trayectorias sociales e identitarias: 150 años de la abolición de la esclavitud en Colombia,* edited by Claudia Mosquera, Mauricio Pardo, and Odile Hoffman, 13–42. Bogotá: Universidad Nacional de Colombia, Instituto Colombiano de Antropología e Historia, Institut de Recherche pour le Développement, Instituto Latinoamericano de Servicios Legales Alternativos.

Múnera, Alfonso. 1998. *El fracaso de la nación: Región, clase y raza en el Caribe colombiano (1717–1810).* Bogotá: Banco de la República y El Ancora Editores.

Múnera, Alfonso. 2006. "María de Jorge Isaacs: La otra geografía." *Poligramas* 25: 49–61.

Napier, A. David. 2003. *The Age of Immunology: Conceiving of a Future in an Alienating World.* Chicago: University of Chicago Press.

Nelson, Alondra. 2002. "Introduction: Future Texts." *Social Text* 20, no. 2: 1–15.

Neubauer, John. 1986. *The Emancipation of Music from Language: Departure from Mimesis in Eighteenth-Century Aesthetics.* New Haven, CT: Yale University Press.

Nieto Olarte, Mauricio. 2000. *Remedios para el Imperio: Historia natural y la apropiación del nuevo mundo.* Bogotá: Instituto Colombiano de Antropología e Historia.

Nieto Olarte, Mauricio. 2006. *La obra cartográfica de Francisco José de Caldas.* Bogotá: Universidad de los Andes.

Nieto Olarte, Mauricio. 2007. *Orden social y orden natural: Ciencia y política en el Semanario del Nuevo Reino de Granada.* Madrid: CSIC.

Nieto Olarte, Mauricio. 2010. *Americanismo y eurocentrismo: Alexander von Humboldt y su paso por el Nuevo Reino de Granada.* Bogotá: Universidad de los Andes.

Nieto Olarte, Mauricio, Paola Castaño, and Diana Ojeda. 2005. "El influjo del clima sobre los seres organizados y la retórica ilustrada en *el Semanario del Nuevo Reyno de Granada." Historia Crítica* 30: 91–114.

Noguera Mendoza, Aníbal. 1980. *Crónica grande del Río de la Magdalena.* Recopilación, notas y advertencias Aníbal Noguera Mendoza; traducciones Otto de Greiff. Bogotá: Ediciones Sol y Luna.

Novak, David. 2008. "2.5 X 6 Metres of Space: Japanese Music Coffeehouses and Experimental Practices of Listening." *Popular Music* 27 (1): 15–34.

Obeso, Candelario. 1877. *Cantos populares de mi tierra* (original copy National Library, 1950 edition, 2005 edition, 2010 edition). Bogotá: Alcaldía Mayor de Bogotá, Fundación Gilberto Alzate Avendaño.

Ochoa Gautier, Ana María. 2004. "Sobre el estado de excepción como cotidianidad: Cultura y violencia en Colombia." In *La cultura en las crisis latinoamericanas,* edited by Alejandro Grimson, 17–42. Buenos Aires: CLACSO.

Ochoa Gautier, Ana María. 2006. "Sonic Transculturation, Epistemologies of Purification and the Aural Public Sphere in Latin America." *Social Identities* 12 (6): 803–25.

Olender, Maurice. 1992. *The Languages of Paradise: Race, Religion and Philology in the Nineteenth Century.* Cambridge, MA: Harvard University Press.

Oliveira Montardo, Deisi Lucy. 2009. *Através do Mbaraka, Música, danca e Xamanismo Gauarani.* São Paulo: EDUSP.

Ortiz, Fernando. [1940] 1987. *Contrapunteo cubano del tabaco y el azúcar.* Caracas: Biblioteca Ayacucho.

Ortiz Mesa, Luis Javier. 2005. "Guerras civiles e Iglesia Católica en Colombia en la segunda mitad del siglo XIX." In *Ganarse el cielo defendiendo la religión: Guerras civiles en Colombia, 1840–1902,* edited by Grupo de investigación religión, cultura y sociedad, Universidad Nacional, 47–86. Bogotá: Universidad Nacional de Colombia.

Otero Muñoz, Gustavo. 1928. *La literatura colonial de Colombia seguida de un cancionerillo popular recogido y comentado por D. Gustavo Otero Muñoz.* La Paz, Bolivia: Imp. Artistica.

Palacios, Marco. 1995. *Entre la legitimidad y la violencia: Colombia 1875–1994.* Bogotá: Grupo Editorial Norma.

Palmié, Stephan. 2002. *Wizards and Scientists: Explorations in Afro-Cuban Modernity and Tradition.* Durham, NC: Duke University Press.

Payzant, Geoffrey. 2002. *Hanslick on the musically beautiful: Sixteen essays on the musical aestheticas of Edward Hanslick.* Christchurch, New Zealand: Cybereditions.

Peñas Galindo, David Ernesto. 1988. *Los bogas de Mompox: Historia del zambaje.* Bogotá: Tercer Mundo Editores.

Pérez, Angela. 2004. *A Geography of Hard Times: Narratives about Travel to South America, 1780–1849*. Albany: State University of New York Press.

Pérez Arbelaez, Enrique. 1959. *Alejandro de Humboldt en Colombia*. Bogotá: Edición de la Empresa Colombiana de Petróleos.

Pérez Silva, Vicente, ed. 1996. *La autobiografía en Colombia*. Bogotá: Imprenta Nacional de Colombia. Digitalized for the virtual library of the Banco de la República in 2004. Accessed February 2010. http://www.banrepcultural.org/blaavirtual/literatura/autobiog/auto5.htm.

Perlman, Marc. 2004. *Unplayed Melodies: Javanese Gamelan and the Basis of Music Theory*. Berkeley: University of California Press.

Picker, John M. 2003. *Victorian Soundscapes*. Oxford: Oxford University Press.

Pidal, Pedro José. 1851. "De la poesía castellana en los Siglos XIV y XV." In *El Cancionero de Juan Alfonso de Baena (siglo XV) ahora por primera vez dado á luz*, xi–lxxix. Madrid: M. Rivadeneyra.

Pineda Camacho, Roberto. 2000. *El derecho a la lengua: Una historia de la política lingüística en Colombia*. Bogotá: Ediciones Uniandes.

Pollock, Sheldon. 2000. "Cosmopolitan and Vernacular in History." *Public Culture* 12 (3): 591–625.

Porqueres i Gené, Enric. 2002. "Nación, parentesco y religión: El nacionalismo vasco entre la sangre y la tierra de la patria." In *Identidades, relaciones y contextos: Estudis d'antropologia social i cultural* 7, coordinated by Joan Bestard Camps, 47–70. Barcelona: Universitat de Barcelona.

Potter, Pamela M. 1998. *Most German of the Arts: Musicology and Society from the Weimar Republic to the End of Hitler's Reich*. New Haven, CT: Yale University Press.

Povinelli, Elizabeth. 2002. *The Cunning of Recognition: Indigenous Alterities and the Making of Australian Multiculturalism*. Durham, NC: Duke University Press.

Povinelli, Elizabeth. 2011. "The Governance of the Prior." *Interventions: International Journal of Postcolonial Studies* 13 (1): 13–30.

Pratt, Mary Louise. 1992. *Imperial Eyes: Travel Writing and Transculturation*. London: Routledge.

Prescott, Lawrence E. 1985. *Candelario Obeso y la iniciación de la poesía negra en Colombia*. Bogotá: Instituto Caro y Cuervo.

Quijano, Aníbal. 2000. "Coloniality of Power and Eurocentrism in Latin America." *International Sociology* 15 (2): 215–32.

Quijano, Aníbal. 2008. "Coloniality of Power, Eurocentrism and Social Classification." In Moraña, Dussel, and Jáuregui, *Coloniality at Large: Latin America and the Postcolonial Debate*, 181–224.

Rama, Angel. [1984] 1996. *The Lettered City*. Durham, NC: Duke University Press.

Rama, Angel. [1984] 2007. *Transculturación narrativa en América Latina*. Buenos Aires: Ediciones El Andariego.

Ramos, Julio. [1989] 2003. *Desencuentros de la modernidad en América Latina: Literatura y política en el siglo XIX*. Santiago, Chile: Editorial Cuarto Propio.

Ramos, Julio. 1994. "A Citizen Body: Cholera in Havana (1883)." *Dispositio* 19 (45): 179–95.

Ramos, Julio. 2010. "Descarga Acústica." *Papel Máquina* 2 (4): 49–77.

Rancière, Jacques. [1998] 2011. *Mute Speech: Literature, Critical Theory, and Politics*. New York: Columbia University Press.

Rancière, Jacques. 2006. *The Politics of Aesthetics*. London: Continuum.

Rath, Richard Cullen. 2003. *How Early America Sounded*. Ithaca, NY: Cornell University Press.

Rehding, Alexander. 2000. "The Quest for the Origins of Music in Germany circa 1900." *Journal of the American Musicological Society* 53 (2): 345–85.

Rehding, Alexander. 2003. *Hugo Riemann and the Birth of Modern Musical Thought*. Cambridge: Cambridge University Press.

Reichel-Dolmatoff, Gerardo. 1991. *Los ika, Sierra Nevada de Santa Marta, Colombia: Notas etnográficas, 1946–1966*. Bogotá: Centro editorial Universidad Nacional de Colombia.

Reill, Peter Hanns. 2005. *Vitalizing Nature in the Enlightenment*. Berkeley: University of California Press.

Restrepo, Eduardo, and Axel Rojas, eds. 2004. Introducción. *Conflicto e (in)visibilidad: Retos en los estudios de la gente negra en Colombia*, 17–32. Popayán: Editorial Universidad del Cauca.

Révész, Geza. 2001. *Introduction to the Psychology of Music*. Mineola, NY: Dover Publications.

Richard, Nelly. 1993. "Alteridad descentramiento culturales." *Revista Chilena de Literatura* 42: 209–15.

Rivera Cusicanqui, Silvia. 2010. *Violencias (re)encubiertas en Bolivia*. La Paz: Editorial Piedra Rota.

Rodríguez-García, José María. 2004. "The Regime of Translation in Miguel Antonio Caro's Colombia." *Diacritics* 34 (3–4) (fall/winter): 143–75.

Rodríguez-García, José María. 2010. *The City of Translation: Poetry and Ideology in Nineteenth-Century Colombia*. New York: Palgrave Macmillan.

Rojas, Silvia. 2004. "Biografía de Rufino José Cuervo." Editado en la Biblioteca Virtual del Banco de la República, 2004–12–07. Accessed September 20, 2009. http://www .banrepcultural.org/blaavirtual/biografias/cuerrufi.htm.

Romero, Mario Germán, ed. 1976. *Epistolario de Ezequiel Uricochea con Rufino José Cuervo y Miguel Antonio Caro*. Bogotá: Instituto Caro y Cuervo.

Rosolato, Guy. 1974. "La voix: Entre corps et langage." *Revue Française de psychanalyse* 38: 75–94.

Rueda Enciso, José Eduardo. 2009. "Esbozo biográfico de Jorge Isaacs." *Revista CS en Ciencias Sociales* 4: 21–53.

Rupke, Nicolaas A. 2008. *Alexander von Humboldt: A Metabiography*. Chicago: University of Chicago Press.

Sa, Lucia. 2004. *Rainforest Literatures: Amazonian Texts and Latin American Culture*. Minneapolis: University of Minnesota Press.

Sabato, Hilda. 2001. "On Political Citizenship in Nineteenth-century Latin America." *American Historical Review* 106: 1290–1315.

Sáchica, Luis Carlos. 1991. "El indígena en Colombia." In *Aspectos nacionales e internacionales sobre derecho indígena*. Serie B, Estudios Comparativos, Estudios Especiales, no.

24, edited by Instituto de Investigaciones Jurídicas, 167–80. México: Universidad Nacional Autónoma de México.

Sahlins, Marshall. 2008. *The Western Illusion of Human Nature: With Reflections on the Long History of Hierarchy, Equality and Sublimation of Anarchy in the West, and Comparative Notes on Other Conceptions of the Human Condition.* Chicago: Prickly Paradigm Press.

Sahlins, Marshall. 2013. *What Kinship Is—And Is Not.* Chicago: University of Chicago Press.

Samper, José María. 1862. *Viajes de un colombiano en Europa. Primera Serie.* Paris: Imprenta de E. Thunot y C.

Samuels, David. 2004. *Putting a Song on Top of It: Identity and Expression on the San Carlos Apache Reservation.* Tucson: University of Arizona Press.

Samuels, David, Louise Meintjes, Ana María Ochoa, Thomas Porcello. 2010. "Soundscapes: Toward a Sounded Anthropology." *Annual Review of Anthropology* 39: 329–45.

Sánchez, Efraín. 1999. *Agustín Codazzi y la Comisión Corográfica de la Nueva Granada.* Bogotá: Banco de la República.

Sánchez Mejía, Hugues R. 2011. "De esclavos a campesinos, de la 'roza' al mercado: Tierra y producción agropecuaria de los 'libres de todos los colores' en la gobernación de Santa Marta (1740–1810)." *Historia Crítica* 43: 130–55.

Sanders, James E. 2004. *Popular Politics, Race, and Class in Nineteenth-Century Colombia.* Durham, NC: Duke University Press.

Sandroni, Carlos, 2010. "Samba de roda, patrimonio imaterial da humanidade." *Estudos Avançados* 24 (69): 373–88.

Santamaría-Delgado, Carolina. 2007. "El bambuco y los saberes mestizos: Academia y colonialidad del poder en los estudios musicales latinoamericanos." In *El giro decolonial: Reflexiones para una diversidad epistémica más allá del capitalismo global,* edited by Santiago Castro-Gómez and Ramón Grosfoguel, 195–216. Bogotá: Pontificia Universidad Javeriana. Universidad Central, Siglo del Hombre Editores.

Santamaría-Delgado, Carolina. 2014. *Vitrolas, rocolas y radioteatros: Hábitos de escucha de la música popular en Medellín, 1930–1950.* Colección Opera Eximia. Bogotá: Pontificia Universidad Javeriana y Banco de la República.

Schmidt, Leigh Eric. 2000. *Hearing Things: Religion, Illusion, and the American Enlightenment.* Cambridge, MA: Harvard University Press.

Schutz, Gunther, ed. 1976. *Epistolario de Rufino José Cuervo con filólogos de Alemania, Austria y Suiza y noticias de las demás relaciones de Cuervo con estos países y sus representantes.* Vols. I and II. Bogotá: Instituto Caro y Cuervo.

Seeger, Anthony. 1987. *Why Suyá Sing: A Musical Anthropology of an Amazonian People.* Cambridge: Cambridge University Press.

Seremetakis, C. Nadia. 1996. *The Senses Still: Perception and Memory as Material Culture in Modernity.* Chicago: University of Chicago Press.

Serje de la Ossa, Margarita. 2005. *El revés de la nación: Territorios salvajes, fronteras y tierras de nadie.* Bogotá: Uniandes—CESO.

Shaviro, Steven. 2009. *Without Criteria: Kant, Whitehead, Deleuze, and Aesthetics.* Cambridge, MA: MIT Press.

Shaviro, Steven. 2012. Blog entry. "Forms of Life." In *The Pinocchio Theory*. June 13. http://www.shaviro.com/Blog/?p=1069.

Sierra Mejía, Rubén, ed. 2002. "Miguel Antonio Caro: Religión moral y autoridad." *Miguel Antonio Caro y la cultura de su época*, 9–31. Bogotá: Universidad Nacional de Colombia.

Silva, Renán. 2003. "El periodismo y la prensa a finales del siglo XVIII y principios del XIX en Colombia." In *Working Documents Universidad del Valle: CIDSE*. Accessed November 2011. http://econpapers.repec.org/paper/col000149/004110.htm.

Silva, Renán. 2005. *República liberal, intelectuales y cultura popular*. Medellín: La Carreta Histórica.

Silva, Renán. 2009. *Universidad y sociedad en el Nuevo Reino de Granada: Contribución a un análisis histórico de la formación intelectual de la sociedad colombiana*. Medellín: Carreta Editores.

Silverstein, Michael. 1993. "Metapragmatic Discourse and Metapragmatic Function." In *Reflexive Language: Reported Speech and Metapragmatics*, edited by John A. Lucy, 33–58. Cambridge: Cambridge University Press.

Sloterdijk, Peter. [1988] 2006. *Venir al mundo, venir al lenguaje: Lecciones de Frankfurt*. Valencia, Spain: Pre-Textos.

Smith, Bruce R. 1999. *The Acoustic World of Early Modern England: Attending to the O-Factor*. Chicago: University of Chicago Press.

Smith, Colin C. 1999. "The Vernacular." In *The New Cambridge Medieval History*, volume 5, c. 1198–c.1300, edited by David Abulaifa, 71–83. Cambridge: Cambridge University Press.

Smith, Mark M., ed. 2004. "Introduction: Onward to Audible Pasts." *Hearing History: A Reader*, ix–xxii. Athens: University of Georgia Press.

Smith, Mark M. 2006. *How Race Is Made: Slavery, Segregation and the Senses*. Chapel Hill: University of North Carolina Press.

Smith, Woodruff D. 1991. *Politics and the Sciences of Culture in Germany, 1840–1920*. New York: Oxford University Press.

Smith-Stark, Thomas C. 2005. "Phonological Description in New Spain." In *Missionary Linguistics II: Orthography and Phonology: Selected Papers from the Second International Conference on Missionary Linguistics Amsterdam Studies in the Theory and History of Linguistic Studies*, vol. 109, edited by Otto Zwartjes and Cristina Altman, 3–64. Amsterdam: John Benjamins.

Solano D., Sergio Paolo. 1998. "De bogas a navegantes: Los trabajadores del transporte por el Río Magdalena, 1850–1930." *Historia Caribe* 2 (3): 55–70.

Steege, Benjamin. 2012. *Helmholz and the Modern Listener*. Cambridge: Cambridge University Press.

Stengers, Isabelle. 1997. *Power and Invention: Situating Science*. Theory Out of Bounds, vol. 10. Minneapolis: University of Minnesota Press.

Stengers, Isabelle. 2009. *Au temps des catastrophes; Résister à la barbarie qui vient*. Paris: La Découverte.

Sterne, Jonathan. 2003. *The Audible Past: Cultural Origins of Sound Reproduction*. Durham, NC: Duke University Press.

Sterne, Jonathan. 2011. "The Theology of Sound: A Critique of Orality." *Canadian Journal of Communication* 36 (2): 207–25.

Sterne, Jonathan. 2012. *MP3: The Meaning of a Format*. Durham, NC: Duke University Press.

Sterne, Jonathan, ed. 2012a. "Sonic Imaginations." *The Sound Studies Reader*, 1–17. Durham, NC: Duke University Press.

Stolze Lima, Tania. 1999. "The Two and Its Many: Reflections on Perspectivism in a Tupi Cosmology." *Ethnos* 64 (1): 107–31.

Stolze Lima, Tania. 2005. *Um peixe olhou para mim: O povo Yudjá e a perspectiva*. São Paulo: UNESP.

Strathern, Marilyn. 1988. *The Gender of the Gift: Problems with Women and Problems with Society in Melanesia*. Berkeley: University of California Press.

Strathern, Marilyn. 1992. *Reproducing the Future: Essays on Anthropology, Kinship, and the New Reproductive Technologies*. New York: Routledge.

Szendy, Peter. 2007. *Sur écoute: Esthétique de l'espionnage*. Paris: Minuit.

Szendy, Peter. 2008. *Listen: A History of Our Ears*. New York: Fordham University Press.

Szendy, Peter. 2009. "The Auditory Re-turn (the Point of Listening)." Paper presented at the conference *Thinking Hearing, The Auditory Turn in the Humanities*, Butler School of Music, University of Texas, October 2–4.

Taussig, Michael. 1993. *Mimesis and Alterity: A Particular History of the Senses*. New York: Routledge.

Thacker, Eugene. 2011. *In the Dust of this Planet: Horror of Philosophy*, vol. 1. Winchester, UK: Zero Books.

Thomas, Downing A. 1995. *Music and the Origins of Language: Theories from the French Enlightenment*. Cambridge: Cambridge University Press.

Tirado Mejía, Alvaro. [2001] 2007. *El estado y la política en el siglo XIX*. Bogotá: El Ancora Editores.

Tomlinson, Gary. 2007. *Singing of the New World: Indigenous Voice in the Era of European Contact*. Cambridge: Cambridge University Press.

Tovar, Antonio. 1986. *El linguista español Lorenzo Hervás: Estudio y selección de obras básicas*. Madrid: Sociedad General Española de Librería.

Triana y Antorveza, Humberto. 1987. *Las lenguas indígenas en la historia social del Nuevo Reino de Granada*. Bogotá: Instituto Caro y Cuervo.

Trouillot, Michel-Rolph. 1992. "The Caribbean Region: An Open Frontier in Anthropological Theory." *Annual Review of Anthropology* 21: 19–42.

Trouillot, Michel-Rolph. 1995. *Silencing the Past: Power and the Production of History*. Boston: Beacon Press.

Tutino, John. 2011. *Making a New World, Founding Capitalism in the Bajío and Spanish North America*. Durham, NC: Duke University Press.

Uribe T., Carlos Alberto. May 13, 2005. "Pioneros de la antropología en Colombia: El padre Rafael Celedón." In digital publication on the web page of the Biblioteca Luis Ángel Arango del Banco de la República. Accessed April 16, 2009. http://www.lablaa .org/blaavirtual/publicacionesbanrepbolmuseo/1986/bo117/boc0a.htm.

Uribe Uribe, Rafael. [1912] 1994. *De cómo el liberalismo político colombiano no es pecado*. Bogotá: Editorial Planeta.

Uricoechea, Ezequiel. 1854. *Memoria sobre las antigüedades neo-granadinas*. Berlin: Librería de F. Schneider.

Uricoechea, Ezequiel. 1860. *Mapoteca colombiana, Colección de los títulos, de todos los mapas, planos, vistas etc. relativos a la América Española, Brasil é islas adyacentes. Arreglada cronológicamente i precedida de una introducción sobre la historia cartográfica de América.* London: Trübner y Cía.

Uricoechea, Ezequiel. 1871. Introduction. In *Gramática, vocabulario, catecismo i confesionario de la lengua Chibcha según antiguos manuscritos anónimos e inéditos, aumentados y correjidos por E. Uricoechea,* edited by Ezequiel Uricoechea, ix–lx. Colección Lingüística Americana, vol. I. Paris: Maisonneuve.

Uricoechea, Ezequiel. 1877. Introduction. In *Vocabulario Páez—Castellano, catecismo, nociones gramaticales y dos pláticas conforme a lo que escribió el señor Eujenio del Castillo i Orozco, cura de Tálaga con adiciones, correcciones y un vocabulario Castellano-Páez por Ezequiel Uricoechea,* edited by Ezequiel Uricoechea, vii–xxiv. Colección Lingüística Americana, vol. II, Paris: Marsonneuve.

Uricoechea, Ezequiel. 1878. Introduction. In *Gramática, catecismo i vocabulario de la lengua Goajira por Rafael Celedón con una introducción i un apéndice por Ezequiel Uricochea.* Collection Linguistique Américaine, 11–48, vol. v. Paris: Maisonneuve Cia, Libreros—Editores.

Uricoechea, Ezequiel. 1976. *Epistolario de Ezequiel Uricoechea con Rufino José Cuervo y Miguel Antonio Caro.* Introduction and notes by Mario Germán Romero. Bogotá: Instituto Caro y Cuervo.

Uricoechea, Ezequiel. 1998. *Epistolario de Ezequiel Uriocechea con Juan María Gutiérrez, varios colombianos y August Friederich Pott.* Edited by Mario Germán Romero. Bogotá: Instituto Caro y Cuervo.

Valderrama Andrade, Carlos. 1997. *Miguel Antonio Caro y la Regeneración: Apuntes y documentos para la comprensión de una época.* Bogotá: Instituto Caro y Cuervo.

Valentine, Elizabeth. 2000. "Carl Stumpf and English Music Psychology." In *Brentano Studien* 9: 251–65.

Vallejo, Fernando. 2007. El lejano país de Rufino José Cuervo. *El Malpensante,* no. 76 (February–March). Accessed July 25, 2009. http://www.elmalpensante.com/index .php?doc=display_contenido&id=323.

Velasco, Daniel. 2000. "'Island Landscape': Following in Humboldt's Footsteps through the Acoustic Spaces of the Tropics." *Leonardo Music Journal* 10: 21–24.

Velásquez Ospina, Juan Fernando. 2012. *Los ecos de la villa: La música en los periódicos y revistas de Medellín.* Medellín: Tragaluz Ed.

Vergara y Vergara, José María. 1867. *Historia de la literatura en la Nueva Granada desde la conquista hasta la independencia (1538–1820).* Bogotá: Imprenta de Echeverría Hermanos.

Viveiros de Castro, Eduardo. 1992. *From the Enemy's Point of View: Humanity and Divinity in an Amazonian Society.* Chicago: University of Chicago Press.

Viveiros de Castro, Eduardo. 1996. "Os Pronomes Cosmológicos e o Perspectivismo Ameríndio." *Mana* 2 (2): 115–44.

Viveiros de Castro, Eduardo. 2004. "The Transformation of Objects into Subjects in Amerindian Ontologies." *Common Knowledge* 10 (3): 463–84.

Viveiros de Castro, Eduardo. 2010. *Metafísicas caníbales: Líneas de antropología postestructural.* Buenos Aires: Katz Editores.

Viveiros de Castro, Eduardo. 2011. *The Inconstancy of the Indian Soul: The Encounter of Catholics and Cannibals in 16th-Century Brazil.* Chicago: Prickly Paradigm Press.

Viveiros de Castro, Eduardo. 2011a. "'Transformaçao' na antropologia, transformaçao da 'antropologia'" *Sopro* 58 (September): 3–16.

Von der Walde, Erna. 1997. "Limpia, fija y da esplendor: El letrado y la letra en Colombia a fines del siglo XIX." *Revista Iberoamericana*, no. 178–79: 71–83.

Von der Walde, Erna. 2007. "El 'cuadro de costumbres' y el proyecto hispano-católico de unificación nacional en Colombia." *Arbor, ciencia, pensamiento y cultura* 183 (724): 243–53.

Wade, Peter. 1993. *Blackness and Race Mixture: The Dynamics of Racial Identity in Colombia.* Baltimore: Johns Hopkins University Press.

Wade, Peter. 2002. *Race, Nature and Culture: An Anthropological Perspective.* London: Pluto Press.

Wade, Peter. 2007. *Race, Ethnicity and Nation: Perspectives from Kinship and Genetics.* New York: Berghahn Books.

Wagner, Roy. [1975] 1981. *The Invention of Culture.* Englewood Cliffs, NJ: Prentice Hall.

Weidman, Amanda J. 2006. *Singing the Classical, Voicing the Modern: The Postcolonial Politics of Music in South India.* Durham, NC: Duke University Press.

Weidman, Amanda J. 2011. "Anthropology and the Voice." *Anthropology News* 52 (1): 13.

Weiss, Julian. 1990. *The Poet's Art: Literary Theory in Castile c. 1400–60.* Oxford: Society for the Study of Medieval Languages and Literatures.

Whitehead, Alfred North. [1927/28] 1978. *Process and Reality.* New York: Free Press.

Yost, William A. 2001. "Auditory Localization and Scene Perception." In *Blackwell Handbook of Perception*, edited by E. Bruce Goldstein, 437–69. Malden, MA: Blackwell.

Yost, William A. 2008. "Perceiving Sound Sources." In *Auditory Perception of Sound Sources*, edited by William A. Yost, Arthur N. Popper, and Richard R. Fay, 1–12. Springer Handbook of Auditory Research Series edited by Richard R. Fay and Arthur N. Popper. New York: Springer.

Zalamea, Fernando. 2009. *América—una trama integral: Transversalidad, bordes y abismos en la cultura americana, siglos XIX y XX.* Bogotá: Universidad Nacional de Colombia.

Zwartjes, Otto, and Even Hovdhaugen. 2004. *Missionary Linguistics, Selected Papers from the First International Conference on Missionary Linguistics, Oslo, 13–16 March 2003.* Amsterdam: John Benjamins.

Zwartjes, Otto, and Maria Cristina Salles Altman. 2005. *Missionary Linguistics, Vol. 2: Orthography and Phonology.* Amsterdam Studies in the History and Theory of Linguistic Science. Amsterdam: John Benjamins.

Zwartjes, Otto, Gregory James, and Emilio Ridruejo, eds. 2007. *Missionary Linguistics, Vol. III: Morphology and Syntax.* Amsterdam Studies in the History and Theory of Linguistic Science. Amsterdam: John Benjamins.

INDEX

acoustemology, 33, 34, 50, 60–62
acoustic: biopolitics, 150, 163; ecology, 65, 75, 208, 212, 219n23; epistemologies of, 3; feedback, 82; inscription of, 25; knowledge, 34; movement, 68; ontologies of, 3; perception, 34; practices, 4; reproduction technologies, 205; sciences, 213; site for analysis, 6
acoustic assemblages, 22–23, 66
acousticity, of the ear in the voice, 205
acoustic object, ambivalence of, 212
Adelung-Vater Mithridates collection, 131
Adler, Guido, 43
aesthesis: of excess, 136–37; of folkloric, 168; of literary, 145; of the local, 27; of popular, 167, 169; of vocal, 166, 169
Afrodescendants, 4, 11, 52, 53, 58; anthropology of, 117; auralities of, 16; burial practices of, 115; construction of, 53–54; identity of, 55–56, 60; Inquisition and, 59; in *María*, 117, 118; musical expression of, 102, 117; musical practices of, 115; occult and, 119; ritual practices of, 61; silencing of, 119; song of, 111; voice of, 121
Agamben, Giorgio, 9
alabao, 115
alarmist tradition, 168, 182
Almario, Oscar, 110, 116, 117, 119
alterity, 63, 64, 110–11; constitution of, 20; orality and, 14; transformation of, 59
Amazon, 29
America: nature of, 143; ontology of, 145; as originary, 136; origin of, 125, 132, 134, 148; as paradisiacal, 124, 135–36, 144

Amerindians: history of, 26; silencing of, 102. *See also* Burskinka; Chibcha; Goajiro; indigenous people; Muisca
Ames, Eric, 46
Andersen, Hans Christian, 1
Anderson, Benedict, 147
Andes, microclimates of, 136
animal: sounding like, 11–12; sounds of, 72–75
animism, 62, 63
anthropology, 81, 135; comparative, 47, 48; equivocation in, 23–24; in Latin America and Caribbean, 25; romantic, 153–54; transductive, 24
anthropotechnology, 17, 19, 20, 172, 210; elocution, 176; temporal, 186; training voice and ear, 203
Arango, Lazaro Mejia, 219n3
archaeology, 153; evolutionary theory and, 156
archaism, 184, 185, 186
Arendt, Hannah, 216n10
Arias, Julio, 54
Aristotle, 18
art, theory of, 49–50
articulate vocality, 170
artist-soldiers, 89
Asad, Talal, 149
audile techniques, 3, 4, 81, 90, 210; of grammarian elite, 108; politics of inscription and, 120. *See also* hearing; listening; mishearing
audiovisual litany, 13–14
audition, occult and, 73
auditory inscription, 82; politics of, 79

Caro, Miguel Antonio, 1, 18, 27, 128, 167, 168, 178, 193, 208; career of, 147–48; climate and race, 158; Cuervo and, 181; *Darwinismo y las misiones*, 27, 125–26, 148–50, 152–53, 156–59, 161–62; faith and science, 156–57; grammar and, 228–29n4; Isaacs and, 27, 125–26, 146–50, 151–53, 157–58, 161–63; *Manual of Elocution*, 201, 207–8; pronunciation and, 174; rhetorical style of, 148–49, 160; speech at Colombian Academy of Letters (1881), 187; tradition and, 158, 160–61

Cartagena, 218n13

Carter, Paul, 50

Castellanos, Juan de, 86

Castro, Vivieros de, 94

Castro-Goméz, Santiago, 13

Catherine the Great, 131, 224–25n6; 225n7

Catholic Church, in public sphere, 88

Catholic government, 173

Catholic Hispanist project, 119

Catholicism, 87, 119; language and, 88; official religion of Colombia, 148, 160; social contract and, 160; in villages/rural regions, 54

Catholic state, 161

Cauca, 145, 219n1, black labor and, 113; description of, 113; free blacks in, 116; Isaacs and, 119; musical practices of, 114; in *María*, 77; Négritude in, 117; population of, 113; slaves in, 116; social makeup, 116

Ceballos, Diana, 59–60

Celedón, Rafael, 140; Goajiro grammar of, 135, 151, 153

census (1777–80), 52–53, 218n14

Chakrabarty, Dipesh, 11

champanes, 35–36

Charles V, 83

Chibchas, 85, 87, 221n9; grammar of, 123, 135–36, 138, 142

Chimila, 150–51

Christian, to "speak in," 119

Christianity, 17, 119, 146; indigenous language and, 126. *See also* Catholic Church; Catholicism

Chua, Daniel K., 102

cimarronaje, 104, 108

citizens, 28

citizen-scholar, 116

citizenship, 88; ethical training for, 203; Spanish language and, 127; speech and 208; theologico-political definition, 127

civilization, vs. barbarism, 173

civil war, 78, 88, 89, 219n2

class, 185

clerical state, 145. *See also* confessional state

climate: character type and, 98, 99; humans and, 175; race and, 12

Cochrane, Charles Stuart, 38

cognitive gestures, 21

Collection Liguistique Américaine (Uricoechea), 123–24, 129–30, 138, 151

collective, 50

collectivity, music production and, 66

Colombia: epistemology of, 2; independence from Spain, 218n16; vs. New Granada, 215n2; originary texts of, 90; as paradise, 2–3; population of, 218n14

Colombian Academy of Letters, 15, 78

Colombian Caribbean, 110; vs. Andean region, 189, 221–22n13; global modernity and, 10–11; indigenous tribes of, 150

colonialism, 13; Anglo settlers and, 11; in Latin America and Caribbean, 9–10, 17; modernity and, 7, 9; resistance to, 56; Spanish, 11; in villages/rural regions, 54

colony: acoustic practices of, 4; histiography of, 215n6; mimetic obsession of, 105; population of, 56; *romance* and, 86; transition to postcolony, 9, 28, 168

comparativism, 24–26, 81, 140, 156; Darwin's theory and, 157; expressive culture and, 26, 152; grammar, 187; indigenous language and, 26; linguistics, 12, 152, 153, 187, 216n13; musicology, 12, 37; philology, 124–25, 130–34, 137, 141–42; problem of meaning and, 142; religious authority in, 152

composition, 212

confessional state, 148–50, 161–63

Connor, Steven, 34, 81, 226n10

conquest, 10, 50, 82, 84, 85, 87; critique of, 152; politico-theological authority of, 125–26

conservatism, 160

conservative government, 78

Conservative Party, 78, 79, 145, 147, 220n4

conservatives, 219n3, 226–27n16; Catholic, 79, 103

Constitution of 1810, 218n16

Constitution of 1886, 147–48, 227n19; Article 35, 160

Conto, Cesar, 119, 145

contrastive listening, 4

conversion, 124; politics of, 162. See also conversos

conversos, 59, 119

coplas, 83, 93, 111, 112, 120, 197; African, 95

Corbin, Alain, 149

costumbrismo, 97, 99, 103, 108–12. See also cuadro de costumbres; expressive culture; folklore

Crary, Jonathan, 216n14

Creoles, 11, 34, 42, 98, 111, 137, 144; anthropology of, 117; Bourbon reforms and, 98; elite, 12, 53, 126; as foreigners, 137; indigenous people and, 154; language of, 104; postcolony, 168

creolization, 110, 187; limits of, 15

Cristina, María Teresa, 112

Crosby, Alfred W., 215–16n7

cuadro de costumbres, 39, 99. See also costumbrismo

Cuervo, Rufino José, 1, 18, 105, 128, 167, 168, 188, 192; Appuntaciones criticas sobre el lenguaje bogotano, 181, 225–26n8; career of, 181–82; Diccionario de construction y regimen de la lengue castillean, 18, 181; work of, 228–29n4

cultural relativism, 63

culture, 12, 92, 190; as act of creation, 65; as artifice, 94; of blacks, 104; imbrication of, 121; natural vs. civilized, 48–49; nature and, 21, 60, 99, 212; religion and, 121; theological, 177

Cunin, Elizabeth, 56, 57; on Afrodescendants, 58

Darwin, Charles: evolutionary theory of, 157; on musical origins, 217n10

Darwinism, 132–34, 146, 155–64. See also evolutionism

Deas, Malcolm, 147, 160

decir, vs. cantar, 90

decolonial: Americanist project, 125, 128–30; anthropology, 24; history of voice, 20–21; theories of, 14–15; thought, 11

denaturalizing, 21

Derrida, Jacques, 9

Diario Oficial, 148

disjuncture, 226n13; between aurality and orality, 138, 144; colonial, 51; linguistic, 139, 141; between mouth and ear, 138; between sign and phone, 140; between Spanish alphabet and indigenous sounds, 137

diversity, Ibero-American struggles, 79

ear, 82; acoustic, 203; as biological instrument, 200–201; conceptions of, 22; dubious, 171, 176, 179, 180, 203; educable, 201; fallibility of, 18; knowledge construction by, 163; musical, 193; physics of, 208; physiology of, 44, 208; purification practices, 163; refunctionalization of, 3; for research, 163; training, 5, 180, 199, 202, 203; transductive, 155; voice and, 7, 8, 28–29, 210, 214

earslip, 81. See also mishearing

echolic mimicry, 51. See also mimesis

echolocation, 50–51, 59

ecomusicology, 67, 212

education, 165

El Cid, 85

Ellis, Alexander J., 43

El Mosaico, 15, 78, 99, 103, 128, 222n20

El Tradicionista, 147

elegance, 152

elite, in Cauca, 116. See also Creole; lettered elite

elocution, 172–79; art of, 208; voice as instrument of, 18

eloquence, 17, 204; as anthropotechnology, 172, 190; constitutive of culture, 190; constitutive of the people, 18–19, 166, 180; governmentality and, 178; objectives of, 176, 190, 204; orthography and, 189; pedagogy of, 168, 179; rhetoric and, 177–78; vulgarity and, 185

encomienda, 52

enemy, 159–60

enlightened vitalism, 49, 71. *See also* Berlin Counter-Enlightenment; vitalism

Enlightenment, 49, 79

enunciation: politics of, 120; practices of, 90

environment: destruction of, 11, 29; music and, 65

envoicing, 61, 63, 64, 65, 170

epic, 86, 93; foundational, 87; indigenous, 86

epistemology: acoustic object and, 212; ontology and, 212–13; of purification, 67

equivocation, 23–26

erasure: of Caribbean Colombia, 10–11; colonial, 54; of indigenous song, 87; of indigenous sources of knowledge, 68–69; in *María*, 118

Erlmann, Viet, 6

Escobar, Tico, 26, 113

Espejo, Miguel de, 85

Esposito, Roberto, 20

Estudio sobre las tribus indígenas del Magdalena (Isaacs), 27, 78, 112, 125, 126, 145–60

ethnography, 26, 112, 146, 153; emergence of, 127; fiction and, 120; *María* as, 120; as transduction, 24

ethnologists, musicians and, 27

ethnomusicologists, 49

ethnomusicology, 43, 44, 45, 47, 212

etymologist, role of, 183

etymology, 17, 18–19, 166, 172, 180–91, 204; constitutive of culture, 190; eugenesis of tongue, 181; nation-state and, 183; patrimonial eugenesic technique, 183;

process of decomposition and recomposition, 188–89; stability of language, 186; temporality, 181; vulgarity and, 185; zoopolitics of selection, 188

eugenesis: of tongue, 19, 181; of voice, 174, 214

European languages, 8

European Romanticism, 79. *See also* Romanticism

Europeans, 11, 34, 42, 67

evolutionism, 133–34, 159; race and, 159. *See also* Darwinism

expansionism, 53

expressive culture, 1–3, 5, 19; Afrodescendant, 118, 178, 183, 186; comparativism and, 26, 152; conventionalized, 91; etymology and, 183, 186; nationalism and, 28. *See also* folklore; poetry; popular expression; song

Fallon, Diego: *Arte de leer, escribir y dictar musica, Sistema Alfabetico*, 191, 193; career of, 191–92; *Nuevo system de escritara musical*, 191; orthographic musical notation of, 19–20, 191–92, 194–99, 201, 203; pedagogy of, 193–94, 197, 198, 201–2

fatherland, 93

Faudree, Paja, 229n1

Feld, Steven, 33

fine arts, vs. popular arts, 5

folklore, 1, 16, 17, 189; collection of, 1–3, 190; disciplinization of, 211; etymology and, 19, 189; immunization of, 205; history of, 79; as literary form, 190; oral, 180; orality and, 166; study of, 77; vocality and, 172. See also *costumbrismo*; expressive culture; folk song; poetry

folkloristics, 166–67, 189

folklorists, 67; documentation by, 120; etymology and, 188–89; idealization by, 166; Isaacs and, 16; rise of, 8

folk song: collection of, 111; documentation of, 121; European Romanticism and, 79; nation building and, 121. *See also* expressive culture; folklore; poetry; popular expression; song

fonocentrism, 17

format, 8, 192

Foucault, Michel, 9, 161

foundational myth, violence and, 159

free people of all colors, 52, 54, 55, 57, 68–69, 121; in Caribbean Colombia, 52; Inquisition and, 59; rebellion of, 56; threat of, 57

French Revolution, 149

galerón, 95

Garrapatas, Battle of, 78

gaze, 6–7

Gene, Porqueres i, 100

genocide, 10, 53

genre, 28, 41, 66–67, 190, 211; aural, 3; Colombian, 2; conquest and, 50; creation of, 26; dispersed, 3; epic, 86; folkloric, 79, 189–90; inscription, 6, 33; literary, 19, 39, 99; mestizaje, 93–94; missionary grammars and, 140; musical, 19, 28, 34–35, 50, 61, 94, 66–67, 114–15, 203, 221n12; national, 93–94; of orality, 80; poetry, 84–85; popular music, 81, 114, racialization, 95; song, 102; soul and, 94–95; Uricoechea and, 129; vocal, 6. *See also* bambuco; *costumbrismo*; *romances*

Geographical Commission, 2

geographic expeditions, 2

geography, 98; character types and, 97; racialized, 113

German Counter-Enlightenment. *See* Berlin Counter-Enlightenment; vitalism

German philology. *See* comparatism: philology

German Romanticism. *See* Romanticism

Gitelman, Lisa, 7

given, the made and, 22, 75

Goajiro: grammar of, 150, 151, 153; orthography of, 140–41

Goehr, Lydia, 101

Goethe, Johann Wolfgang von, 71

Gomez, Adolfo Leon, 160

Gomez-Bravo, Ana María, 221n12

Gonogoza, Helcias Matan, 117

Goodman, Steve, 217n3

Gosselman, Auguste, 37–38

governance of the prior, 186, 187, 190. *See also* politics of the prior

governmentality, 163, 165–66, 178, 203, 213

governmentalizing, 212

Gramatica, vocabulario, catecismo i confesionario de la lengua Chibcha segun antiguos manuscritos anonimos e ineditos, aumentados y correjidos por E. Uricoechea (1871), 123, 135–36, 138, 142

grammar, 18, 167, 179; aesthetics of, 125; authority of, 190; of Bello, 164; of Celedón, 151, 153; Chibcha, 123, 135–36, 138, 142; collection of, 130–31, 224–25n6; colonial, 2, 16, 26; comparative, 124, 133, 153, 187; disjuncture in, 137–39; Goajiro, 151, 153; of Hervas y Panduro, 130–31, 142; history of American continent and, 125, 144; human origins and, 224n1; indigenous, 27, 124, 125, 130, 132, 136–38; of Isaacs, 125, 145–46, 150–55; missionary, 124, 131–32, 135, 136, 138, 139, 140, 152, 154–55, 226n11; music and, 175, 177; nation-state and, 189–90; of Obeso, 78, 103–4; political use of, 79; priestly, 146, 151; religious use of, 125–26, 162; as training, 167, 174; of Uricochea, 27, 123–26, 128–30, 132–44, 224n1; of vocality, 6, 167

grammarian presidents, 2, 20, 27, 79, 147, 160, 227n18. *See also* Caro, Miguel Antonio; Marroquín, José Manuel

grammarians, 105, 108–9, 125, 160, 177, 181, 187. *See also* grammarian presidents

grammarian state, 125, 221–22n13, 228–29n4

grammatilization, 167

Grimm, Jacob, 1, 44

Gutiérrez, Juan María, 127, 129, 134

poetry of, 92; population of, 61; public sphere and, 144, 149, 164; religious conversion of, 124, 153, 162–63; reorganization of, 61; resistance by, 55, 56; songs of, 75, 82, 93; sounds of, 43; subjugation of, 52; voice of, 11–12, 28. *See also* Amerindians; Burskinka; Chibchas; Goajiro; Muisca

indigenous social movement, 29, 163

indigenous studies, 11

indigenous texts, 7

inoculation, 171

Inquisition, 54, 59–60, 84, 119, 163, 219n18

inscription, 3, 25, 33; acoustic stereotypes, 103; alternative, 104, 105; aural, 6, 7; bodily, 212; of colonial and postcolonial, 3, 40–41, 135; genres of, 6; lettered, 90; linguistic, 138–41, 155, 184, 192; model for the popular, 166; of music, 191–205; politics of, 16, 27, 33, 79, 81, 90–121, 125–27, 155, 211; practices of, 23, 58, 80, 111–12; sites of, 7; of song, 81–82, 88–121; of sound, 7–8; technologies of, 7–8, 90, 111, 172, 190; vs. transcription, 104–5. *See also* anthropotechnologies; eloquence; etymology; orthography

Isaacs, Jorge, 15, 77–78, 93, 103, 108, 111–20, 125, 154, 155; anthropological interests, 113, 125; career of, 16, 27, 78, 112, 116, 119, 145, 146; Caro and, 27, 125–26, 146–50, 151–53, 157–58, 161–63; as citizen-scholar, 115–16; as convert, 119, 152–53; death of, 224n29; ethnographic sensibility of, 26–27, 112, 145, 146, 150, 153–54, 155–57; exile of, 163; Jewish heritage, 16, 78, 119; military career of, 89; as poet, 117, 163; politics of hearing, 150–51; radical liberals, 79, 88–89, 112, 119–20, 145; *La revolución radical en Antioquia*, 112, 145; secular humanism of, 148–50, 153, 159; songbook of, 112. See also *Estudio sobre los tribus indígenas del Magdalena*; *María*

Isabel la Católica, 220n7

Jáurequi, Carlos, 108–9; 223n23

Jews: Antillean, 152–53; *conversos*, 59, 117, 119; Dutch, 152; in *María*, 117, 118–19

Jones, William, 124

Juan II, 84

juga, 114–15

jugulars, 85

Julian, Antonio, 226n9

juntas, 60

juruna, 62

Keane, Webb, 97

kulturvölker, 47, 61

La Araucana, 86

Lacoue-Labarthe, Philippe, 88, 89, 140, 202

Lafuente y Alcántara, Emilio, 222n16

Landers, Jane, 219n17

language: body and, 185; ethnic descent and, 124; families, 142–43; genealogy of, 184; as living body, 180, 188; music and, 197, 211, 229n1; nation-state and, 182, 185, 208; phylogenesis of, 185; purification of, 163; politics of, 8, 162; popular culture, 8; spoken, 184; temporalization of, 183–84; uniformity of, 174; vs. voice, 174; written, 175, 184. *See also* indigenous language; Spanish language

Las Casas, Bartolomé de, 11

Latin Americanism, 168

Latour, Bruno, 209

Lazarus, Moritz, 47

Leclerc, Georges Louis, 49

LeGuin, Elizabeth, 223n22

LeMoyne, Auguste, 40–42

letrados, 163

lettered city, 5, 16, 29, 88, 107, 177; blood purity and, 119; citizenship in, 88; immunization and, 177; indigenous language and, 149, 163; Isaacs and, 119; Obeso and, 15–16, 104–10; orality and, 166; political theology of, 145, 163

modernity, 10, 13, 209–10; voice and, 166

mohán, 86

Mompox, 15, 78, 103; a population of, 52; as port city, 31, 52; rebellion, 218n12; travel to, 36

Morales, Guillermo Abadia, 112

Moreno, Jairo, 21, 22

Mosquera, Claudia, 53

mouth: disjunction with ear, 137–38; as instrument, 193; orality and, 140; pronunciation and, 138, 140, 174; training, 180; written text and, 7

Muisca, 82, 86, 156, 221n9, 227n21

Muller, Max, 1

multiculturalism, 94

multinaturalism, 63

Muñoz, Gustavo Otero, 1, 77

music: anthropology of, 61–62, 120; art vs. folk, 211; aurality of, 80; cultural, 47; definition of, 49; disciplines, 43–44; dissemination of, 19, 190; grammar and, 175; instrumental vs. vocal, 102, 200; as language, 197, 203; language and, 102, 197, 211, 229n1; making, 65–66; mimesis and, 66–67; as moral force, 20, 201–3; as multiplicity, 67; natural, 34–35, 47, 114; orality and, 211; origins of, 42, 45–47, 217n10; orthographic transcription of, 19–20, 191–92, 194–99, 201, 203; personhood and, 17; physical effect of, 45; popular, 100; praxis, 66; as science, 199; sound and, 65, 75; theological transcendence of, 177; traditional, 34–35, 112; ventriloquization of, 198; Western art, 45, 47, 100, 101; world, 102

musicalization, 193

musical notation, 19–20, 191–92, 194–99, 201, 203

music history, 210

musician, professionalization of, 191

musicology, 12, 37, 43, 45–47, 49, 81, 102, 212

music theory, 21–22, 43–44, 101–2, 193, 199

Muslims, 59, 119. See also *conversos*

Mutis, José Celestino, 98, 136, 224–25n6

mythology, 156, 157, 162; vs. religion, 162

Nancy, Jean-Luc, 88, 140, 202

nationalism: Basque, 100; collective fratricide and, 89; Colombian, 109; Hispanic, 80; indigeneity and, 28; musical, 12; vernacular and, 147

nation-state: comparativism and, 26; indigenous language and, 19, 25–27, 92; indigenous people in, 155, 186; Magdalena River and, 35; personhood in, 17, 28, 146; political theology of, 25, 155, 212; politics of the prior and, 19, 186, 229n5; proper language of, 17, 28; sovereignty of, 19, 186; vocal pedagogy of, 212. See also clerical state; confessional state; postcolony

natural history, 83; development of, 37, 178

naturalvölker, 47, 61, 110

nature: abundance of, 25–26, 135–36, 144, 154; awe of, 25, 69–71; Berlin Counter-Enlightenment/vitalism and, 48, 71, 203; contested, 11; culture and, 9, 12–13, 20–21, 26, 27, 48, 60–62, 67, 75, 94, 99, 116, 135, 144, 161, 208, 212–14; exuberance of, 108, 110; the given as, 94, 99; human nature and, 49; inscription of, 7; knowledge of, 59; laws of, 48, 70–71; listening to, 68–69, 72–75; local, 3; music and, 33, 44–45, 47, 49, 114; ontology of, 145, ordering of, 4, 129, 202–3, 222n7; philosophy of, 70; race and, 95, 99; as reality, 94; science of, 2, 36–37, 62, 68; silence of, 74–75; sounds of, 40, 41, 47, 50, 67, 137; voice and, 3, 36; voice of, 74. See also denaturalizing; multinaturalism

naturphilosophie, 25–26

negation, 118–19

Négritude movement, 103, 117

neologism, 185

Neubauer, John, 101

New Granada, 31, 39, 49, 82–83, 98, 124, 129, 132; apostles in, 156; botanical expeditions, 98, 136; civil war in, 89–90; conquest of, 83; economic potential of, 136; foundational epic, 85–87; indigenous peoples in, 126; mestizos in, 57–60, 158; nature in, 2, 98; originary texts of, 85, 90; personhood in, 119; poetry of, 91; popular song of, 92–95, 129; population of, 53, 56; racial fuzziness in, 53; Spain and, 7, 83, 85; term, 215n2

New Granada: Twenty Months in the Andes (Holton), 39

newspapers, 191

Nieto, Juan José, 189

Núñez, Rafael, 146–48

Nuskilusta, 46

Obeso, Candelario: Afro-futurism of, 111; audile techniques of, 16; career of, 15–16, 78–79, 103–4; lettered city and, 15–16, 104–10; misspellings of, 104; nature and, 108; orthography of, 15–16, 77, 79, 103, 104–6, 108–9, 110, 223n24; poetry of, 15, 77, 103, 104, 105, 107–8, 223n25; 223–24n26; race and, 104, 108–9; radical liberal, 79, 88; transcription of, 105; as translator, 111. See also *Cantos populares de mi tierra*

occult, 71, 84, 119. *See also* religion; witchcraft

ocularcentricism, 13, 16–17, 24

ontology: of acoustics, 22; American, 134–45; of ear and voice, 8–9; epistemology and, 212, 213; of Latin America, 127; of voice, 205, 208

oral: aural and, 208; governmentality and, 211; history of, 209; spectral dimension of, 204

orality: alterity and, 14; audibility of, 168; aurality and, 140; communal expressivity, 172; definition of, 190; history of, 167; idea of, 121; fidelity and, 204; grammatilization of, 167; as heritage,

186; human vs. nonhuman, 167; lettered city and, 166; music and, 211; the people and, 211; political subject and, 17, 172; political theology and, 15, 20; the popular and, 166; race and, 15; resilience of, 166; in song, 120; studies of, 6; as tradition, 14–15

oratory, 177. *See also* rhetoric

origins: Adamic, 132–33, 134, 158; of America, 134; animal, 44; human, 44; mythology and, 137; of words, 187–88. *See also* Darwinism; etymology; evolutionism

orthography, 19, 166, 172, 204; arbitrary, 104; auralization of, 192; creative misunderstanding of, 90; definition of, 179; dubious, 165, 188; immunization and, 172; of indigenous language, 4, 8, 126–27, 137–43, 155; as musical notation, 19, 191–204; nonstandard, 103; pedagogy of, 17, 179–80; personhood and, 18; the popular, 166; process of uniformization, 174, 189, 192; pronunciation and, 123, 124, 173–74; standardization, 181, 184, 189–91; as technology of acoustic inscription, 90, 190–91

orthology, 166

Ortiz, Fernando, 4

other, 14–15, 17, 21, 100, 110, 205

Our Father. *See* Lord's Prayer

Palacios, Marco, 89

Palenque de Matuderé, 219n17

palenques, 54–56

Pallas, P. S., 131

panopticon, 149

Pardo, Mauricio, 53

Patria Boba, 218n16

patrimony: cultural, 19; etymology and, 183; of orthography, 124; voice and, 188

pedagogy, vocal, 178–79

people, the, 14, 18, 19, 20

personhood: alterity and, 64; construction of, 183; contested nature of, 29;

historical political vs. philosophical juridical, 157; idea of, 5; of indigenous people, 127, 146–47, 149–50, 161–62; juridical, 204; moral definition of, 97; music and, 17; in nation-state, 17, 28, 91, 146; in New Granada, 119; ontological understanding of, 92; orthography and, 18; philology and, 147; political definition of, 11, 28, 149; race and, 12, 54, 92; racialization and, 53–54; silencing of, 92; voice and, 170

perspectivism, 63, 65

philologist, indigenous language and, 153

philology: comparative, 90, 124, 130–31, 132–33, 141; German, 134, 137; grammarian presidents and, 79; indigenous personhood and, 147; Latin American, 132; political power and, 2, 27; Spanish, 81, 84

phonetics, racialized, 151

Pidal, Pedro José, 220nn7,8

Piedrahíta, Lucas Fernández de, 87

pitch, 12, 43–46, 67, 194–95

poetry: Afro-Colombian, 117; as colonizing enterprise, 83; documentation of popular, 121; as gift from blacks to whites, 96; epic, 86, 87, 93; erudite, 84, 85–86; foundational, 90; medieval, 84, 220n8; orality of, 80; pedagogical use of, 180, 197; popular, 15, 77, 82, 84, 85–86, 91–92; race and, 92–93; Renaissance, 79, 85; uncultured, 85. *See also* expressive culture; folklore; folk song; popular expression; song

political theology, 25, 125–27, 148–50, 153–55, 163, 208, 212. *See also* clerical state; confessional state

politics of the prior, 117, 229n5. *See also* governance of the prior

Pollock, Sheldon, 83

polyphonic sonorities, 45

popular, the, 71, 105, 220n5

popular expression: systemization, 167; truth value of, 187. *See also* expressive culture; folklore; folk song; poetry; song

Portuguese, 10

positivism, 217n10

postcolony, 9, 28, 125; citizen-scholars of, 116; Creole elites and, 168; early years of, 80, 110, 165; political-theological authority in, 125; race and, 121; traditional in, 91

Pott, A. F., 225–26n8

Povinelli, Elizabeth, 110

primitives, 70

primitivism, vs. civilization, 70–71

pronunciation: Afro-Caribbean, 103; alternative, 104; bad/improper, 5, 97; of indigenous language, 137–38; orthographic inscription of, 97; problem of, 123, 138, 176; standards, 179

psychoacoustics, 44, 200

psychology, folk, 47

Pythagoreanism, 101–2

race: climate and, 158; construction of, 94–95; geography, 97, 100; idea of, 99; mixing of, 18, 159; orality and, 15; personhood and, 12, 54, 92; and the popular, 93; power and, 159; song types, 91

racialization: of culture, 25; of documentation, 120; of expression, 98; hybridity and, 57; personhood and, 53–54; popular song and, 95; of voice, 121

racism, 107

radical liberalism, 78, 116, 219n3, 226n16; vs. conservative liberalism, 112, 116, 147; in Cuaca, 117, 119, 145. *See also* Isaacs, Jorge; Obeso, Candelario

Rama, Angel, 4, 8, 88, 165

rancherías, 55

Rancière, Jacques, 43, 177–78

rebellion, 56

Regeneration, 78, 147, 173

Regime of Terror, 218n16

Rehding, Alexander, 45–46, 47, 217n10

religion: culture and, 121; imbrication of, 121; invention of the world, 162; vs. myth, 162; vs. occult, 84; in public sphere, 162; vs. science, 154–55. *See also* Catholic Church; Catholicism; Christianity; confessional state; Jews; Muslims; occult

religious purification, 152

representation, politics of, 51

resistance, 60

Restrepo, Eduardo, 54

rhetoric, 27, 79, 90, 129–30, 149, 159–60, 177–78

rhythm, 44

romances, 85, 86, 93, 95

Romanos, Larra and Masonero, 99

Romanticism, 88, 101, 102, 109–10

Sahlins, Marshall, 62, 64, 67

Samper, José María, 89

Santander, Francisco de Paula, 189

Sapir-Whorf hypothesis, 142

Saussure, Ferdinand de, 126, 143, 226n13

science: and language, 161; vs. religion, 154–55, 161; secular authority in, 152

secular humanism, 148–49, 152, 155, 159, 161

secularization: of indigenous affairs, 146; of voice, 121

Seeger, Anthony, 61

segmentation, 82

self: construction of, 64; politics of, 110; as transpersonal, 64

Sepúlveda, Juan Gines de, 11

silence: in *María*, 118; of nature, 74–75; vs. noise, 75; orality and, 120; religion and, 117; untamed vocalities and, 167

silencing: of indigenous people, 97, 102, 121; inscription of, 120; of Jewishness, 16, 119, 121; of personhood, 92; techniques of, 119

Simmel, George, 45, 110

sin, 227n20

slave: as economic figure, 183; escaped, 54–55. See also *cimarronaje*

slavery, 93; abolition of, 110, 116, 119; effect on blacks, 98; history of, 10; as labor system, 52, 56; music and, 223n22

slave trade, 53, 55

Smith, Woodruff D., 47

Smith-Stark, Thomas, 139

Sociedad de Naturalistas Neo-granadinos, 27, 128

song, 15, 77–121; aurality, 80; of blacks, 92; burial, 115; carrier of mass music, 103; classification of, 91; collection of, 79; dissemination of, 7–8; epic, 93; epistemological production of, 80; vs. erudite poetry, 86; as gift, 96; heritage of, 95; history of, 120; inscription of, 7, 79, 82, 102–3; *juglares* vs. troubadours, 221n10; listener and, 22; malleability of, 80; nation and, 100; orality of, 80; physical attributes of, 81; spectrality of, 8; vs. speech, 45; spoken vs. sung, 221n12; theory and, 90; transduction of, 80. *See also* expressive culture; folklore; folk song; poetry; popular expression

songbook, novel and, 112

sonic effect, 34

souls, making of, 94

sound: of animals, 39–40, 61, 65, 72–74; body and, 32, 33; ecology of, 50; fusion of, 217n8; history of, 21; human vs. nonhuman, 5, 51; inscription of, 4, 7, 33; local, 34; as military technology, 214; music and, 65, 75; of nature, 40, 41, 47, 50, 67, 137; neurobiological processes of, 214; vs. noise, 174; orthographic manipulation of, 78; physics of, 208; political theology of, 166; sense-making and, 34; spectrality of, 8, 14, 84; technology, 207

sound studies, 6, 207–9, 213, 214

Sound Studies Reader, 209

Spanish Academy of Language, 189

Spanish language: as "Christian," 119; of Christianity, 17, 83; citizenship and, 127; as conquering, 92; cultural patrimony of, 19; genre and, 93; imperial, 83–84; indigenous language and, 92; of nation-state, 17; as original, 87; pronunciation of, 228n2; standardization of, 179; vulgar vs. Castilian, 186–87

Spanish Reconquest, 88

speech: common, 184; literary, 184; national unity and, 189; proper (*el bien decir*), 185; as representation, 177; vulgar, 184, 185
standardization, 190
states, political division of, 91
Steege, Benjamin, 22, 201
Steinthal, Heymann, 47
Stengers, Isabelle, 29, 213
Sterne, Jonathan, 13, 14, 21, 192, 209, 210, 216n16
Strathern, Marilyn, 50, 64
Stumpf, Carl, 45, 46–47, 217n10
subaltern, 12–13, 14, 57, 61
Szendy, Peter, 6

Taussig, Michael, 42–43
temporal perception, 82
texts, 7
theological-political. *See* clerical state; confessional state; political theology
theory, 90; invention, 89–90; of language, 143; literary as, 140. *See also* literary theory; music theory
Tomlinson, Gary, 42
tonal system, 43. *See also* harmony
tonpsychologie, 44
Torgue, Henry, 33–34
tradition, 91–93, 157–63; modernity and, 166; orality and, 14–15; race and, 93; temporalization of change, 185; transmission of, 135
transcription: audile technique of, 210; of indigenous language, 126–27; of music, 19; of Obeso, 15–16, 104–5
transculturation, 60
transduction, 23–29, 67, 155–64, 216n16
translation, 155
transportation, 35, 56–57
travel writing, 36
Trouillot, Michel-Rolph, 10–11

Uribe Uribe, Rafael, 227n20
Uricoechea, Ezequiel, 105, 111, 145, 151, 224n1, 226n9; on America's origins,

126, 127–35; *Arte y vocabulario de la lengua Chiquita*, 139; career of, 26–27, 127–29; *Diccionario de voces de historia natural americanas* (1873), 129; exile of, 163; grammars of, 27, 123–26, 128–30, 132–44; on human origins, 224n1; orthography of indigenous language, 123–24, 138. *See also* Collection Liguistique Américaine; *Gramatica, vocabulario, catecismo i confesionario de la lengua Chibcha segun antiguos manuscritos anonimos e ineditos, aumentados y correjidos por E. Uricoechea* (1871)

Valladolid, 11
Vatican, 148
verbal art. *See* orality
Vergara y Vergara, José María, 220–21n8, 222n20; on bambuco, 93–94; on black poetics, 96, 99; career of, 15, 78; Catholicism of, 87–88; on epic poetry, 86–87; on foundational poetics, 85–86, 88, 90; literary history and, 82–83, 87; on Muisca, 86–87; nature vs. culture, 93–94; on popular song, 91–103, 120; prologue to *María*, 118; on pronunciation, 97; racialization of, 15, 91, 94–98, 99–100, 120; on Spanish as national language, 91, 92; tradition, 92–93. See also *Historia de la literature en Nueva Granada desde la conquista hasta la independencia* (1538–1820)
violence, 159
vision, 13–14
vitalism, 12, 203
Viveiros de Castro, Eduardo, 11, 23–24, 222n15; nature vs. culture, 94
vocality, 6, 166, 214; articulate, 170
vocalization, practices of, 209
voice: ambiguity of, 204; ambivalence of, 210; of animals, 65; Aristotelian ideas, 228n1; audibility of, 167; as audile technique, 212; of blacks, 97–98; body and, 5, 167, 169, 209; defining culture,

205; disciplining of, 9, 165, 175, 203, 204, 210, 211; dissemination of, 191; ear and, 9, 28–29, 210; elocution and, 175; eugenesic project of, 174; grammatilization of, 18, 167–71, 177, 204; historiography of, 207, 214; human vs. nonhuman, 5, 63, 169–70, 209; immunization of, 17, 20, 171–72, 177; inscription of, 212; instrument of political transformation, 210; vs. language, 171, 174; local sounds and, 5; as machine, 193; mediality of, 209; modernization and, 208; music and, 97–98, 177, 200; nature, 3–4, 61, 74; ontology of, 208; organ of listening, 198; pedagogy of, 204; of the people, 5, 171–72; personhood and, 170; physiology of, 174, 193; as political subject, 9, 165, 171–72; political theology of, 167; politics of, 28, 103, 189–90; popular in, 166; racialization of, 97–98, 121, 175; secularization of, 121; as soul, 102; time and, 166; transpersonal self and, 12, 64–65; untamed, 172; *zoè* and, 9, 170; zoopolitics of, 9, 170, 174, 183, 205, 208. *See also* envoicing

völkerpsychologie, 44–45, 47, 48
Von der Walde, Erna, 99, 148

Wade, Peter, 99
Wagner, Roy, 50–51, 94
Weiss, Julian, 84
Whitehead, Alfred North, 161
whites, 54–55, 59, 93, 97
witchcraft, 59–60. *See also* occult
women, 117
work concept, 20, 65–66, 81, 101–2, 190, 192, 199, 203, 210–11
working class, 58
writing, 6, 8
written city, 165, 177

zambos, 52, 68
zoomusicology, 212
zoopolitics: of person, 173, 204; of selection, 190; of voice, 9, 170, 174, 183, 205, 208, 210
zoè, 9, 170